Critical Studies of Education

Volume 8

We live in an era where forms of education designed to win the consent of students, teachers, and the public to the inevitability of a neo-liberal, market-driven process of globalization are being developed around the world. In these hegemonic modes of pedagogy questions about issues of race, class, gender, sexuality, colonialism, religion, and other social dynamics are simply not asked. Indeed, questions about the social spaces where pedagogy takes place—in schools, media, corporate think tanks, etc.—are not raised. When these concerns are connected with queries such as the following, we begin to move into a serious study of pedagogy: What knowledge is of the most worth? Whose knowledge should be taught? What role does power play in the educational process? How are new media re-shaping as well as perpetuating what happens in education? How is knowledge produced in a corporatized politics of knowledge? What socio-political role do schools play in the twenty-first century? What is an educated person? What is intelligence? How important are socio-cultural contextual factors in shaping what goes on in education? Can schools be more than a tool of the new American (and its Western allies') twenty-first century empire? How do we educate well-informed, creative teachers? What roles should schools play in a democratic society? What roles should media play in a democratic society? Is education in a democratic society different than in a totalitarian society? What is a democratic society? How is globalization affecting education? How does our view of mind shape the way we think of education? How does affect and emotion shape the educational process? What are the forces that shape educational purpose in different societies? These, of course, are just a few examples of the questions that need to be asked in relation to our exploration of educational purpose. This series of books can help establish a renewed interest in such questions and their centrality in the larger study of education and the preparation of teachers and other educational professionals.

More information about this series at http://www.springer.com/series/13431

John A. Weaver

Science, Democracy, and Curriculum Studies

 Springer

John A. Weaver
Georgia Southern University
Statesboro, GA, USA

Critical Studies of Education
ISBN 978-3-030-06729-8   ISBN 978-3-319-93840-0   (eBook)
https://doi.org/10.1007/978-3-319-93840-0

This Springer imprint is published by the registered company Springer Nature Switzerland AG
The registered company address is: Gewerbestrasse 11, 6330 Cham, Switzerland

For Peter Appelbaum, my first academic friend and the most creative person I know
For Annette and Noel Gough, innovative thinkers of science and curriculum studies
For Jimmy Geiger, will you remember his name?
To the judgmental and unread

Acknowledgments

Thank you Shirley Steinberg for all you do for the field of curriculum studies. You are a consistent and dedicated scholar to the field and my work. I am thankful for Springer Publishers for their dedication and hard work on this book especially Marianna Pascale, Truptirekha Das Mahapatra and Suresh Kumar. I am always truly grateful for my scholarly colleagues who understand the importance and value of books in an anti-intellectual culture: Marla Morris, Ming Fang He, Sabrina Ross, Daniel Chapman, and Michael Moore.

Contents

Chapter 1
Science, Democracy, and Curriculum Studies: Why (Not) Science Matters?

> *Instead we should make use of our security, our seniority, to take risks, to make noise, to be courageous, to become unpopular.*
>
> —Carolyn G. Heilbrun, Writing a Woman's Life, 131.

> *Science and Art, Objectivity and Subjectivity, Nature and Society...We spend so much of our intellectual efforts trying to reconcile one to the other, when in fact they may have never been separate in practice.*
>
> —Phillip Mirowski, The Effortless Economy of Science, 47.

> *But where in philosophy, and even outside philosophy...have people pictured thought as other than a path? In it, [A Derrida seminar] Heidegger "invents" an aphorism on Greek thought: 'A path...is never a method.*
>
> —Philippe Lacoue-Labarthe, Poetry as Experience, 108.

1.1 Vignette One: The General Economy of Anything Creative

Early in my academic career I came across this passage from Georges Bataille's (1991, 21 & 25) *Accursed Share*: "The living organism...receives more energy than is necessary for maintaining life; this excess energy (wealth) can be used for the growth of a system...if the system can no longer grow, or if the excess cannot be completely absorbed in its growth, it must necessarily be lost without profit; it must be spent, willingly or not, gloriously or catastrophically....Woe to those who,... insist on regulating the movement that exceeds them the narrow mind of the mechanic who changes a tire." Does this not provide a definition for so much in education and research? Education is treated like a tire and the researchers are mechanics following steps. Yet so much of what is found in the realm of education, like any dimension of life, is an abundance of energy that allows a creative system such as the human mind to grow. To try to manipulate, limit, or artificially harness that energy is to woefully, ingloriously, and catastrophically destroy that system.

© Springer International Publishing AG, part of Springer Nature 2018
J. A. Weaver, *Science, Democracy, and Curriculum Studies*, Critical Studies of Education 8, https://doi.org/10.1007/978-3-319-93840-0_1

Does this not serve as an apt appraisal of public education and those neoliberals who premeditatively and unimaginatively murdered an important institution for a vibrant democracy?

It was, and remains so, Bataille's quote that drove me to explore the general economy of hip hop, popular culture, science fiction, posthumanism, and now science matters. The sciences, like other arts, is a general economy of energy when it is not tightly controlled by neoliberal thought and, its co-conspirator, utility. It is no surprise to me that Arkady Plotnitsky's work on Bohr and Derrida would find Bataillé's notion of a general economy important for understanding the sciences. Bohr's notion of complementarity and Derrida's deconstruction are examples of a general economy. A general economy "must relate to irreducible losses in representation and meaning in any interpretative or theoretical process... This determination introduces a fundamental indeterminacy, a kind of structural 'vagueness' or 'more-or-less-ness' within all (non)systems it considers" (Plotnitsky 1994, 20). I am not alone in seeking out general economies. I think this is a fundamental dimension of the life of a curriculum scholar and can be found in the work of many curriculum scholars. However, I think curriculum scholars often ignore one area where a general economy of energy exists: science. I intend to demonstrate some of the ways science is a vibrant general economy of thought and why curriculum scholars need to pay more attention to this general economy before neoliberalism and utility sap this energy and numb science to its need for creativity.

1.2   Vignette Two: The Arts

In a letter written to Anne Stevenson in 1964 the poet Elizabeth Bishop (Giroux and Schwartz 2008, 861) described a part of her trip to England this way: "I do admire Darwin! But reading Darwin, one admires the beautiful solid case being built up out of his endless heroic *observations*...one *feels* the strangeness of his undertaking, sees the lonely young man, his eyes fixed on facts and minute details, sinking and sliding giddily off into the unknown. What one seems to want in art, in experiencing it, is the same thing that is necessary for its creation, a self-forgetful, perfectly useless concentration." What Bishop admired most in Darwin she also saw in art. Both the sciences and arts were intellectual endeavors in which poets, physicists, biologists, historians, philosophers, astronomers all wandered into the unknown, sharing a common moment of willfully falling into a state of absorption thereby creating an experience that is simultaneously found in the mundane (observations) and uniquely different (strangeness). These experiences when found in the sciences and arts are completely "useless concentration[s]" as time and body are lost (self-forgetful). How is it Bishop is able to see what a scientist and a poet do as similar? When she visited England Darwin's house was a destination listed as a necessity. It seems what drew her to Darwin was his technique, style, in creating his science which to her was his art. Bishop was not mesmerized by science nor interested in transferring its mythical qualities (pure objectivity and impersonal neutrality) to the arts. She did however see the work of science as important to understand life and the arts. This last sentence of course could be written in reverse. There was no competition for her. One need not choose between the arts or the sciences and if one chose unwisely then one was not a philistine or beholden to neoliberal reductionism or utility. As she noted, when done well poetry and science as techniques or styles were strangers to usefulness. What struck me in this passage was the admiration Bishop held for Darwin because he was able to capture the art of his craft. To capture a craft one would have to pay a steep price of loneliness and painstaking attention to fine details but it lead to a giddiness, an experience of intellectual bliss. Bishop ended her paragraph to Stevenson with this: "(In this sense it is always 'escape' don't you think?)" An escape from what? No doubt an escape from time because the arts and sciences allowed one to forget. I think it is an escape from much more. I think Bishop would join C.P. Snow in rejecting the "two cultures" that were fossilizing in European and USA societies and emerging in universities. Arbitrary boundaries between the sciences and arts did not keep Bishop away from Darwin. She found a way to unite with her kindred spirit and recognize his abilities to create something meaningful and lasting. What boundaries will we create as curriculum scholars? Will we choose to escape the utility of time and seek out our kindred spirits in all fields of knowledge?

1.3   Vignette Three: Philosophy

Ever since he studied philosophy while in prison serving a sentence for armed robbery, Bernard Stiegler has become one of the most important contemporary thinkers. In an essay on "Technoscience and Reproduction" Stiegler (2007, 29) lays out a critique of a major shift that has taken place in the last 40 years. Since Aristotle "technics…belongs to the field of contingency, to which is opposed the necessity of science. Such a point of view is obviously incompatible with the very notion of technoscience, whose name indicates the collusion of technology and science." Like Bishop, Stiegler recognizes the beauty of technics. A technique is a contingency in which the creator is pushed into a state of uncertainty. Just because one is entering into a laboratory, creating a model, constructing a poem, writing a novel, recording a history, thinking a thought, or entering a new culture does not mean something will come of it. To think otherwise is not to enter the abyss of contingency. Instead, it is to enter the mythical realm of certainty where experiments follow faithful protocols, poems receive "museical" blessings, histories preordained zeitgeists, novels whimsical inspirations, and cultures paradigmatic shifting gestalt experiences. All of these are qualities found after the fact in rewritten histories that explain away the contingent and turn the creators into something mythical and godlike. They, however, are not found in technics. This striving for certainty is not found in science either. As a result, Stiegler suggests that the opposite of technics is not certainty but it is the necessity of science. This does not mean that somehow science is utility and technics is art and therefore science is inferior to the arts. Science is also a technique, but it also has a quality of necessity to it that makes it different from technics. The problem arises when the differences between the contingency of technics and the necessity of science are erased and a technoscience is formed that places the need for certainty and utility above all else. This erasure creates problems for both art and science. First, art is subsumed under technoscience in the name of utility. Second, science is reshaped. Science under the banner of technoscience represents what Stiegler calls (2007, 31) an "overturning…where *it is science which becomes applied technology, and not technology applied science*. Science as applied technology produces formalized results which become du-plicable, i.e. *reproducible by automations*." The algorithmic supersedes the rhythmic transforming life into an act of reproducible acts done without thought or contingency and done with mindless certainty. This overturning does not create just a crisis for science. We all became algorithmic with the rise of technoscience. As curriculum scholars ignored the sciences we all became technoscientific agents. The merger of technics and science into technoscience created a crisis for "the education system as an apparatus of transmission and reproduction of knowledge, to the point of threatening it to collapse, this technoscience-fiction being itself an absolute revolution of the question of transmission" (Stiegler 2007, 39). Teachers are now judged by the algorithmic as standardized test scores dictate what is important during the school day and students became testing sites for future economic gain, both personal and corporate. Science in universities became STEM and the arts became STEAM while philosophy became history and the rest including history became fodder for 30 second commercials describing how excellent university x is, and where you can get your degree by never setting foot on its physical campus nor experience any life disruptions from classroom demands like reading and writing.

The fault of curriculum scholars is they have blamed the algorithmitization of society on science and fail to see the connections between the contingency of technics or the arts and the necessity of science. As a result they have failed to see demands of certainty placed on everything within what we call generically life. "Technoscience expressly intimates to us the question of knowing what we want because the fiction that reason would today be *forced to project*, as technoscience, becomes the fiction of a *science that no longer betokens the real, but that which INVENTS the possible*. It is still and always a question of invention (and this is why the technosciences from now on are more interested in patents than 'discoveries'), and of the *possibility of adoption*" (Stiegler 2007, 40). Even here I suspect curriculum scholars are prone to target technology as the culprit, naming it as a dehumanizing blight on individuality. But this is not the case. The threat of technoscience is found just where Stiegler finds it. The problem is not technology nor science, it is the "overturning" the necessity of science into a demand in the name of profit. In technoscience science is not a result of a wandering curiosity that Darwin demonstrated in Bishop's eyes, it is the patents any discovery can generate in order to adopt science in the name of corporate start-ups and innovative entrepreneurs. Science is no longer necessary because it could lead to a Galilean, Darwinian, or Pasteurian moment, but because it is the site of future profits, debits, credits, and stock market shares. This transformation of science from a necessity for knowledge to a necessity for profit is what curriculum scholars lost sight of when they refused to see how the sciences are connected to the arts.

1.4   Vignette Four: Anthropology

It is difficulty to discuss any aspect of science or modernity without discussing the work of Bruno Latour, I will not wait any longer. In his recent book, *Facing Gaia: Eight Lectures on the new Climate Regime* Latour (2017, 4) writes "It is impossible to understand what is happening to us without turning to the sciences…And yet, to understand them, it is impossible to settle for the image offered by the old epistemology; the sciences are now and will remain from now on so intermingled with the entire culture that we need to turn to the humanities to understand how they really function." Why do so many curriculum scholars ignore Latour's declaration? Should we not be involved in debates around climate change, global warming, the Anthropocene, the sixth extinction moment? I ask curriculum scholars the same question Latour asks of religious fundamentalists why are you ignoring these matters of science? There is not a second planet to turn to! Many curriculum scholars may object to me equating them to fundamentalists or they might suggest they are interested in our environmental crisis. But as Latour notes one cannot engage in these matters of grave concern without engaging in the sciences and the scientists who create the models. One may be interested in ecocidal tendencies within capitalism, but the level of interest is defined by the level of scientific engagement. To ignore science is to ignore the planet. Yet this is not the most important part of the Latour's comment. The goal of curriculum scholars should not be to turn to scientists and genuflect. Scientists are not the kings and queens of the environment or the universe. They are partners with everyone else on this planet. The key to Latour's comment is that he recognizes that scientists cannot do their important work without the humanities. It is the humanities that will connect science to meaning. When scientists come to curriculum scholars we should not rebuff them, mark off our disciplinary territory, and proclaim a peace treaty as long as they stay on their side of the arbitrary line of knowledge. When, if, they come we should embrace them and present our creative minds to the problem at hand and merge with scientists in order to create an art called science. Will we?

1.5   Vignette Five: Curriculum Studies?

This vignette is yet to be created. This is why I am writing this book, to replace this question mark with a proposal, a call to action hopefully. Will curriculum scholars accept it? I have my doubts. In the field of curriculum studies there is a great deafness of being heard. We pretend to enter into dialogue with other scholars, teachers, students, and citizens throughout the world yet so few curriculum scholars seem to listen to one another let alone other people. Instead there is an opting out of the world, proclaiming technology the problem of the world or utilizing technology for narcissistic proclamations (isn't this a definition of facebook?) and tweeting out feigned fits of rage and victimhood? What safe place on earth are we opting out to? In this book I propose five different ways curriculum scholars can enter into discussions over science matters: rhetoric of science, the sociology of expertise, feminist science, postcolonial thought, and Nietzschean philosophy. Our silence has held us for too long, and the chapters that follow are ways I seek to speak up and out. Will I be heard? This is a question I cannot answer nor will I try to. It is out of my control.

In working on my next project I read Philippe Lacoue-Labarthe's *Poetry as Experience*. In this book Lacoue-Labarthe (1999, 25) discusses Hölderlin's withdrawal from poetry and his decline into madness, and concludes that "[m]adness is, indeed, the absence of artistic production." What shall we name it when curriculum scholars withdrawal from science and refuse to share its artistic productions and abilities with those who could benefit from our artistry? There are examples of curriculum scholars who study various fields within the sciences including chaos and complexity theories, but I would like for curriculum scholars, especially those who distant themselves from science matters, to go further. When science matters are broached it usually is as metaphor or dealing with the question of how can science inform the thinking of curriculum studies. This is fine work, but I would like curriculum studies to engage with/in science and help shape and define the research agendas and policies of the sciences. This is my first goal of this book.

One of the many benefits of writing this book was re-engaging with David Blades' work as I explain in more detail in interlude five. In his important work *Procedures of Power & Curriculum Change: Foucault and the Quest for Possibilities in Science Education* (1997, 1) Blades asks: "Where is hope…perhaps if our children understand the nature, activities, benefits and dangers of science and technology, and the relationship between science, technology and society (STS) the next generation may find a hopeful future." I do not share David's optimism because the forces of neoliberalism are too tightly gripped around the imaginations of many USA citizens, but the sentiments are appreciated. The second goal of this book is to find ways in which non-scientists can get involved in science matters. As I explain in more detail in Chap. 2, I have believed since the early 1990s that science is too important to leave to the scientists. This proclamation requires a reciprocal relationship between scientists and non-scientists. It requires scientists to end their elitist encirclements that they use to demarcate between disciplinary partners and allies and uninformed non-scientists who just do not understand what the "masterminds" are doing. Scientists have to become better pedagogues and educate non-scientists. At the same time, non-scientists have to become better students. That is, non-scientists have to educate themselves first and "read-up" on science matters, find out which scientific topics interest them the most, and then seek out the scientists and engage them in discussions about the importance of the science and how it shapes society. Curriculum scholars specifically have two important tasks at hand. We need to become the educators of the scientists. How is it scientists can best communicate their ideas to non-scientists, and here I do not mean to philanthropists, government officials, venture capitalists, and journalists? This obviously they do (too?) well. I am referring to citizens who need to understand the impact science matters have in shaping their lives beyond economics. In order for curriculum scholars to become the educators of scientists, we have to become immersed in science matters. We have to become what Heidegger referred to as becoming a philosopher without being a philosopher. We need to understand what is the state of knowledge within the fields of science and what are the cutting edge problems facing those fields. I hope some of the readers will take up this cause.

A final goal of this book is to (re)connect or (re)strengthen the ties between science and democracy. As I note in chapter 5, "The Economics of Science, Neoliberal Thought, and the Loss of Democracy", after World War Two scholars like Thomas Kuhn were alarmed that non-scientists would become more involved in science matters so they created the idea that science and democracy were naturally connected and there was no need for non-scientists to be involved in science matters. This hypothesis has been challenged since then and the challenge of neoliberalism to a vibrant democracy and the coopting of science in the name of neoliberal technoscience has demonstrated that there is nothing naturally tying science to the idea of democracy. The tie has to be mutually constructed and vigilantly guarded. I try to demonstrate how this can be done by more non-scientist involvement in science matters.

Besides what is to follow in this section of the introduction I will not try to present any notion of what a democracy is or should be. I do this for two reasons. The first is I do not wish to play the academic game of gottcha. If I provide a detailed definition of what I think a democracy should mean then someone will come along and write "but aha you have forgotten this or your discussion of point A dismisses the importance of point B." I am more interested in living in a democracy then trying to define exactly what a democracy is or might be. This is why I am focusing on science matters. Science is being coopted by neoliberal thought and being used against the rights of individuals, usually in the name of corporate rights never in the name of non-human animals and the vibrancy of an intellectually, culturally, and politically free world. Second, I am a Trotskyite and a Derridean thinker, that is I believe in the need for perpetual revolution. I do not wish to create a definition of what a vibrant democracy might look like and then say well all boxes are checked therefore we live in a democracy. We can all go home now, watch Monday Night Football and drink some beer. Defining democracy and then proclaiming its existence is not an ideal for equality but merely a recipe for complacency. I prefer Derrida's notion of a democracy to come. What is it we should be striving for that exists only on a horizon? What is it we need in order to live up to our ideals of equality and justice for all? Clearly we are not even close to such an ideal. There are too many innocent Black men murdered in our streets, too many poor people without any rights, too many transgendered, lesbian, bisexual, and gay people mocked and discriminated against, too many immigrants harassed because they are marked as illegal and too many "illegal" immigrants terrorized by nativists and xenophobes, and too many non-Christians discriminated against because they believe differently or have the audacity to not believe at all. Have I left any group of people out besides the self-proclaimed Christian "victims" from the "war" on Christmas or the wealthy who proclaimed they are victimized by burdensome tax laws? Of course I did and this is why I wish to not focus on a definition beyond this statement section and instead focus on living a democratic life. I do not wish to become complacent but wish to help those who deserve to live a democratic life, including animals and objects, and then move on to the next revolution, horizon. Complacency is another name for death and I rather put that off for another day, ask me tomorrow what I think and I will tell you the same thing until neither one of us is around to speak. That, however, is for another day.

One last note. I will not introduce each chapter in this introduction like I normally do for my books. Instead I have written interludes. I decided to do this because not only did I want to present to you five different ways one could approach science matters but I wanted to present to you different ways to write about science. Curriculum scholars exist within a field of knowledge that is stifled by a lack of imagination, too many in education think a path for thinking is a method. Method is a path; a path to stagnation. If statistics or quantitative approaches to research do not dominate then qualitative approaches do. This binary thinking stifles creativity and renders educational research helpless against the destructiveness of neoliberal thought, utility, and technoscience. In this book I provide you with four different ways to write about science matters: Vignettes, traditional essays, interludes, and aphorisms. If these are not sufficient enough for your creative appetites please add to this. I welcome your creative additions.

References

Bataille, G. (1991). *Accursed share* (Vol. 1). New York: Zone Books.

Blades, D. (1997). *Procedures of power & curriculum change: Foucault and the quest for possibilities in science education*. New York: Peter Lang Publishers.

Giroux, R., & Schwartz, L. (Eds.). (2008). *Elizabeth Bishop: Poems, prose, and letters*. New York: Farrar, Strauss and Giroux.

Helbrun, C. (1988). *Writing a woman's life*. New York: Norton.

Lacoue-Labarthe, P. (1999). *Poetry as experience*. Stanford: Stanford University Press.

Latour, B. (2017). *Facing Gaia: Eight lectures on the new climate regime*. Cambridge: Polity Press.

Mirowski, P. (2004). *The effortless economy of science*. Durham: Duke University Press.

Plotnitsky, A. (1994). *Complementarity: Anti-epistemology after Bohr and Derrida*. Durham: Duke University Press.

Stiegler, B. (2007) Technoscience and reproduction. In *Parallax 134* (pp. 29–45)

Chapter 2
From Kuhn to the Economics of Science: Curriculum Studies and Science Studies

This chapter is 27 years in the making. When I was a Ph.D. student at the University of Pittsburgh the degree program I was in, comparative education, only required students to take three classes, an introduction to administrative and policy studies, an introductory research class, and a class dealing with education, culture, and society. Beyond these three courses we were free to take any classes we wished. In the class on education, culture, and society the professor, Eugenia Potter, required us to read Thomas Kuhn's *The Structure of Scientific Revolutions*. This book, as most people know, is without a doubt the most influential academic work published in the twentieth century. Before the turn of the millennium, everyone was fascinated with the 100 of the greatest everything. There was even 100 of the greatest non-fiction works published in the twentieth Century. Kuhn's book was the only academic book to make the list. All the other choices were popular non-fiction books written by non-academics. When Eugenia Potter had us read Kuhn's book in 1991 I knew nothing about it and never had a class on the history and philosophy of science. This all changed once I opened *The Structure of Scientific Revolutions*. After reading it, I immediately came to the conclusion that science was too important to leave to scientists. Everyone had to be involved with the making of science and more importantly the making of science policy. The next semester I signed up for an independent study with Professor Potter on the philosophy of science, and I have not stopped reading in the fields of science studies since.

This is my first systematic attempt to explain why it is important for all citizens of a democracy to be involved in science. Too many times we hear policy makers, politicians, and sometimes even science advocates suggest that Science (now so affectionately referred to as STEM or Science, Technology, Engineering, and Mathematics) is essential for everyone to study because of the important economic benefits these fields of knowledge produce. The assumption is that science is important because it has become an engine of innovation and creativity and therefore by extension a motor for economic growth. This, by no means, is my argument. In fact, the fetishizing of science as an economic engine has seriously jeopardized the important connections between science and democracy, science and the humanities,

© Springer International Publishing AG, part of Springer Nature 2018
J. A. Weaver, *Science, Democracy, and Curriculum Studies*, Critical Studies of Education 8, https://doi.org/10.1007/978-3-319-93840-0_2

and science and society. Science has been reduced, like almost everything in the United States especially, to a measure of economic exchange. As a result the important political, cultural, and intellectual roles science plays in society are ignored. My plans for this section are to provide a review of my journey into science studies. Science studies include not only the fields of history and philosophy of science (two fields that often have intense intellectual disagreements regrettably) but also the anthropology and sociology of science plus feminist and postcolonial thought. I will discuss the major works that have shaped these fields and also shaped my thinking on matters of science, education, culture, and democracy. I will do this in order to provide curriculum studies scholars a broad introduction to this field of knowledge in order to eventually demonstrate why science studies is essential for curriculum studies. My other major goal is to eventually progress to the growth of what is called the economics of science. It is my contention that the economics of science is one way for curriculum studies to enter into debates concerning science matters. When I write about the economics of science I have two specific, but separate meanings in mind. The first deals with the actual economic aspects of science that shape how research is done. For instance, this notion of the economics of science often refers to issues such as who funds scientific research and how this funding shapes broad scientific goals, the impact of economic thought on how knowledge is viewed as a commodity as if it were just another object with extrinsic, marketable value, the rise of patents and technology transfer agreements in influencing the exchange and flow of information, the rise of entrepreneurial science inside and outside of the university setting, the hagiography of science and the dominance of an economic myopia used to justify policy decisions within the university, and a whole gambit of issues facing scientists and society in general. The second meaning pertains to the intellectual growth out of and in tandem with science studies that actually looks at the historical, rhetorical, philosophical, and anthropological impact of economic thought on how scientists and society think about the meaning and importance of science. This includes important debates over the coupling of democracy with science and the very real threats to this common law marriage, the rise of economic metaphors to explain developments in science, the dominance of economics to justify the sciences' role in society, and the role of lay people or citizens in the making of science policy and the development of research agendas. These are issues I will cover in the chapters to follow. In this chapter I want to cover the history and philosophy of science that leads up to my coverage of the economics of science and why these fields of knowledge are important for curriculum studies scholars to commit their intellectual energies.

2.1 A Kuhnian Revolution Without Kuhn

Not only because my forays into the history and philosophy of sciences began with a reading of Thomas Kuhn's work but mainly because his work is such a watershed point in the understanding of science, I will begin my discussion with Kuhn's *The*

Structure of Scientific Revolutions. Written in 1962 and revised twice, Kuhn's work is often cited in almost every field in the social sciences and the humanities. In spite of its mythical status, there is much about the book that is ignored, but needs to be remembered both as a contribution to the history and philosophy of science and as a period piece. While earning credit for starting what we refer to now as science studies, later in his life Kuhn renounced much of the work done in his name. He was a reluctant revolutionary whose primary goal was not to open up new fields of study but rather to justify the autonomy and importance of science in a democratic society. In this sense, it is best to read Kuhn's work in the same vein as Immanuel Kant's (1979) *The Conflict of the Faculties*, John Henry Newman's (1996) *The Idea of a University*, and C.P. Snow's (1998) *The Two Cultures*. What these four works have in common is a deep desire to protect their fields of knowledge (in Kant's case philosophy, Newman's religion, Snow and Kuhn's science) from all intrusions, be they industrial, political, democratic, or secular. As Philip Mirowski points out (2004, 67) one of Kuhn's primary interests in writing is to suggest that "science has nothing to do with the layman's understanding of himself or of society. Therefore, the lay populace should simply acquiesce in the revolutions and upheavals that rock the sciences…and continue their funding and obedience." The lay person should just trust scientists and accept that which the scientist creates and the changes that ensue. This perspective is not something espoused by a revolutionary, but rather someone interested in protecting the rights and privileges of scientists no matter what type of society they might function in and no matter what the demands of the lay person may be. From this perspective one can begin to see why Kuhn might have rejected much of what came after his work in the field of science studies. Talk about lay person involvement in science policy and science involvement in industrial developments that sparked major social, economic, and cultural transformations could jeopardize the importance of scientific autonomy. It was better for Kuhn to perpetuate the myth of scientific neutrality and objectivity than to focus on the deep connections between society and science.

This desire to maintain the autonomy of scientists can go a long way to explain also why Kuhn placed so much attention on theory as a driving force for practice. It is this approach that could pave over the realities of scientific involvement in social sectors such as the military, industry, and of course government, and the influences these sectors played in shaping scientific agendas. Kuhn assumed that it was theory that drove scientific developments and eventually revolutions that would change how science was done and what research scientists would conduct. As a result all other aspects of science were ignored including the role instruments, or as Bruno Latour would say non-human actants, play in the development of science, and, of course, as mentioned the role non-scientific institutions such as industry, political entities, non-government agencies, and heaven forbid lay people might play in the development of scientific agendas. In Kuhn's version of science, the scientist is not a revolutionary, although his (and in Kuhn's eyes and so many others, the scientist is gender specific) ideas may be. Instead the scientist goes about conducting normal science or science that follows a set protocol that cannot be ignored. As a result of following the rules of science, the scientist may introduce a new idea, discover a new component of a theory thereby adding a new piece to the puzzle. Normal science is almost universally accepted by everyone I will discuss in this chapter. As the French Sociologist Pierre Bourdieu will note to not follow accepted protocol's is to risk banishment from the field of knowledge. Accepted protocol is one of the major ways in which individual scientists are acculturated into their field of knowledge.

Yet, in spite of the set, explicit and implicit, rules of conducting scientific work, events happen in the laboratory or in nature that normal science cannot explain. When scholars in other fields discuss the importance of Kuhn's work they often ignore this important process and jump right to paradigm shifts and this is what Kuhn is most noted for in fields such as curriculum studies.

In 1890, Max Planck was studying Black Body radiation and he noticed that Quanta when radioactive act in unpredictable ways. This uncertainty, however, could not be explained by Newtonian laws, and Planck, being the good and faithful scientist he was, immediately assumed he was in error. He dismissed his findings as an anomaly and an outlier. This for Kuhn is a common occurrence in science: anomalies are dismissed and the status quo accepted as set and firm law. For 10 years Planck found this anomaly to be true and eventually came forward with well documented evidence that Newtonian laws could not function in the atomic realm and as a result began the Quantum revolution in science. As Kuhn noted a Gastalt switch in the mind is thrown and a new way of seeing the world is founded, new agendas are established, new stars in the field anointed, awards presented to the most creative and influential, new disciplinary fields of knowledge are founded, and sometimes whole fields of industry are established. Before all this can happen, however, a struggle ensues in which scientists struggle for authority within their field. Note that in this struggle, in Kuhn's mind, there are no politicians pushing one agenda over another, industries trying to tilt the debates in their favor, or university presidents getting involved. This struggle is a private affair, untainted by outside forces, and truth prevails eventually unmolested by non-scientists' motives. In Kuhn's work scientific paradigms reach a point of incommensurability in which the paradigms do not know how to communicate with one another. They adopt different research agendas, ask different questions, create different methods, and seek to undermine the others' claims of authority to name truth and science. Because these paradigms are so different there is no way to independently judge which paradigm is The Truth, the struggle to establish scientific authority demonstrates how theories are laden with assumptions of truth and beliefs about what it is a scientist does. It is this point that many scholars will focus on and demonstrate how notions of truth and political, economic, and societal beliefs shape science thereby destroying the whole notion of a scientist being value neutral. Eventually what happens, Kuhn notes, in this struggle between incommensurable paradigms is the older paradigm proponents eventually die. As a result of the reality of time, the victorious paradigm's representatives get to educate the new scientists using their techniques, methods, and ideas, influence the funding agencies to adopt their research agenda, and more importantly, they get to write the textbooks and tell why their paradigm is the truth and the way. The story of science then never becomes a story of unrelenting progress in which one paradigm went as far as it could go and another one picked up where the other left off. Science becomes a story in which one paradigm is overtaken by another, research in one area stops, not necessarily completely but dramatically, and another one begins. Both paradigms produce results, but each produce different results. These different results demand different ways of doing science, and different ways of seeing the world. In spite of what I now see as major flaws in Kuhn's ideas concerning scientific progress, when I initially came across his work it was exactly what I wanted. I wanted to know how science worked, how knowledge was constructed, how scholars struggled to name and define truth, and how academic work was given meaning in the creative and imaginative process of everyday interactions.

After I read Kuhn's work, I began to focus on the philosopher of science Paul Feyerabend. Feyerabend (1988, 14) is someone who immediately pigued my attention when in the opening line of his book *Against Method* he wrote "The idea of a method that contains firm, unchanging, and absolutely binding principles for conducting the business of science meets considerable difficulty when confronted with the results of historical research." I always believed that education researchers were obsessed with methodological questions. A finding, conclusion, hypothesis, or fact was not correct unless it was created or discovered using the perfect methodological protocol. When I moved from a history Ph.D. program to comparative education I heard people often use the phrase historical method. I thought what is that? We never talked about method, historians talk about historiography and interpretation. Feyerabend feed those suspicions I held towards the dominance of methodology in education. Feyerabend's thesis is that if science depended solely on the scientific method to make truth claims and discoveries, the world would still be in the Stone Age. When it comes to making truth claims about scientific matters, for Feyerabend, anything goes. To make his argument he used the case study of Galileo trying to convince the Church leaders that his support of a Copernican view of the solar system is correct and the Church leaders approach to science needs to be revised. The debate between Galileo and The Roman Catholic Church is often told as a battle between science and religion. It is not. It was a debate between two alternative ways of conducting scientific research and how to interpret the results. The Catholic Church represented the established Aristotelian science protocol of its day and the philosophy that underpinned it while Galileo was the rebel, rebuking traditional science and offering an alternative. To support his perspective Galileo was not a completely rational thinker who did not embellish anything with rhetoric, supplying only accepted, pure facts to draw carefully vetted conclusions. Feyerabend (1988, 67) argues that "Galileo uses propaganda. He uses psychological tricks in addition to whatever intellectual reasons he has to offer." He also uses conjecture, suggesting that the reason the Church scientists cannot see what he sees through his telescope is because their eyes are tricking them and they must reject what they see with their naked eye because light bends while travelling through space. It will not be until another 400 years before Galileo's belief is verified as true. Politics play an important role in science as well. Galileo knew the matter was not about understanding Truth, it was about challenging the Roman Catholic

Church and no matter what Galileo thought his telescope revealed about the stars he was going to lose the debate because, no matter how many arguments he presented to support Copernicus' perspective it was not enough to convince the Church scientists. In the politics of science, like all political realms, power wins, at least temporarily. It was not as if Galileo lost the rhetorical and political debate because he did not know how to use these tools of persuasion. Feyerabend establishes that he indeed knew how to debate scientific points. Mario Biagioli (1993) demonstrates that Galileo knew very well how to play the political game of science. Today to be a successful scholar in any field one needs to create something of note, demonstrate its importance, gain notoriety because of the creation (this can be awards such as the Nobel Prize, media appearances, or acceptance as a leader in one's field of knowledge), secure peer acceptance of one's priority over other possible creators, and, more than any other time, secure patents or copyright and possible monetary gain. In Galileo's time being a successful scholar was very different. The key of the game in the 1600s was to secure a patronage from a powerful political leader. For Galileo, the patron of choice was Cosimo de Medici. Galileo's audience was not, therefore, peers who would review his evidence, weigh his claims, and pronounce the importance of his findings. Cosimo II was his audience. Galileo had to convince the Medici family of the worthiness of his claims because any discoveries, inventions, or creations would bring honor and fame to the Medici family and the more honor and fame they received the faster Galileo would climb up the patronage ladder as a favorite scientist of a wealthy family who would support his future efforts. When using his telescope Galileo discovered the moons of Jupiter he presented them as gifts to the Medici family and referred to them as the Medicean stars. This earned Galileo a favorite status for future scientific endeavors. Biagioli (1993, 105) notes that the Galilean gift was not accepted because "of their technological usefulness or scientific importance, but" they were "prized" as "spectacles, as exotic marvels." The Medici recognition allowed Galileo to leave the university and pursue his own research agenda. Galileo also sought the patronage of the Church and he thought he had sufficiently developed it when he began to challenge Church scientific doctrine. This was his miscalculation. As Biagioli (1993, 297) suggests: "From Galileo's point of view, the publication of the *Assayer* [Galileo's polemic challenging Aristolelian science as it applied to comets] marked a patronage clash between two networks he had carefully developed over the years." Although Galileo's moves were methodical in the sense of being very well planned out, the courtier politics he played in order to gain respect and acceptance of his research agenda demonstrates Feyerabend's point that anything goes in scientific arguments and there is not one set way of doing science. As was the case with Kuhn, Feyerabend's observations on science were exactly what I was looking for; an indepth exploration of scientists at work.

Next, I decided to go more abstract in my thinking about science and started to read the French sociologist, Pierre Bourdieu. While Kuhn and Feyerabend's work was always focused on science, Bourdieu's was not. As a sociologist his interest in science focused on how economic, social, and cultural capital translated in every-day interactions between scientists and society. The historian of science Timothy

Lenoir (1997, 15) succinctly captures Bourdieu's interest in science when he writes: "The scientific field is a field of positions occupied by agents with differential stances toward one another. Each field, whether politics, economics, art, literature, or science, has its own logic. To play for stakes in the scientific field requires a specific form of capital, such as educational experience and appropriate material resources...it also includes instruments of circulation, such as journals or publishing houses, which choose to publish articles and books in accordance with certain criteria...Each field also has institutions concerned with consecrating good work through, for instance, bestowing prizes and membership in academies." This description of science as a field of knowledge is not different from other fields such as curriculum studies except the sciences may have more political, social, and cultural capital. But the amount of capital a field of knowledge holds always depends on the circumstances. For instance, there are rare occasions where curriculum studies might possess more capital than the sciences that can include policy decisions concerning public education or in developing understandings of how race, gender, sexual identity, ethnicity, social class, and religion shape the lives of people in all institutions. There is no doubt that if the topic is identity politics and formation, curriculum studies wields more capital than the sciences. No matter what the circumstances may be Lenoir's description holds true. Scholars are interacting with one another in order to control and define what the meaning and purpose of a field of knowledge may be; one's educational experience (such as who one studies with matters) and material resources (such as where one studies, what one's rank is, and what one's access to resources is) matter. It also matters who controls the journals, book series, and the flow of ideas, and the jockeying for prizes and membership. It happens in every field. In all of these struggles and maneuverings, "the specific issue at stake," for Bourdieu (1975, 19), "is the monopoly of scientific authority, defined inseparately as technical capacity and social power, or to put it another way, the monopoly of scientific competence, in the sense of a particular agent's socially recognized capacity to speak and act legitimately...in scientific matters." Those scientists who are able to prove competence, which can take many forms including following accepted protocol but also when one has accumulated enough authority to construct at a minimum a slightly different protocol that can support one's research agenda and support one's findings so scientific facts can be established. "Because all scientific practices are directed towards the acquisition of scientific authority," Bourdieu (1975, 21) suggests, it is important to understand how this struggle takes place and how authority is secured. Establishing priority is the most important way in which a scientist establishes authority. Once the community of scholars recognize that a scientist established priority for a discovery of a new phenomenon, a new technique, or a new research line of work, then that authority is put to work by "imposing a definition of science implying that genuine science requires the use of a great scientific bureaucracy provided with adequate funds, powerful technical aids and abundant manpower; and they present the procedures of large sample surveys, the operations of statistical analysis of data, and formalization of the results, as universal and eternal methodology, thereby setting up as the measure of all scientific practice the standard most favourable for their personal and institutional capacities" (Bourdieu 1975, 21). This game of establishing authority is not dictatorial. One trying to establish authority needs to appeal to the community of scholars as peers and judges of the validity of a certain research agenda and approach. "The struggle for scientific authority," Bourdieu (1975, 23) believes, "… owes its specificity to the fact that the producers tend to have no possible clients other than their competitors…only scientists involved in the area have the means to symbolically appropriating his work and assessing its merit." When trying to assert one's scientific authority one must balance the need for priority and uniqueness and established rules of protocol with a field. To be too cautious risks gaining any authority at all and never adding to a field of knowledge new understandings or new techniques and to go too fast in establishing authority can risk ostracism from the field. How much a scientist should risk to gain authority depends on one's position within the community. For instance when Edwin Schrödinger wrote *What is Life* in 1944 he took a risk leaving the field of physics and writing about what he thought was the new frontier of science research: Biology. Schrödinger took the risk by leaving his field in which he won a Nobel Prize and contributed greatly to and entering another field with established authorities and protocols. However, Schrödinger with his personal authority as established by his work in Physics and a Nobel Prize was not taking that great of risk compared to someone who was just entering a field without an established reputation. It is with junior scholars that the move to gain scientific authority plays out most intensely. The call for any new person within any field of knowledge is to contribute to the field either through new interpretations, discoveries, methods, techniques, or lines of research, but to do any of this is to risk expulsion from the field. As Bourdieu (1975, 23) notes "the scientist who appeals to an authority outside the field cannot fail to incur discredit." Yet, the new scientist is expected to challenge the very status quo that may very well discredit them. For Bourdieu new scholars face two choices as to how they wish to structure their careers. They can attempt a road of research that moves them "towards risk-free investments of succession strategies [accepted research protocols]" but give them "limited innovations within authorized limits," or they can adopt "subversion strategies, infinitely more costly and more hazardous investments which will not bring them the profits accruing to the holders of the monopoly of scientific legitimacy unless they can achieve a complete redefinition of the principles of legitimating domination" (Bourdieu 1975, 30).

2.2   An Anthropologist Studies Modernity and the Objects It Creates

To this point the focus on science has been about the importance of theory and how it shapes what scientists create. This changes with the arrival of Bruno Latour and his anthropological studies of science. Instead of focusing on the macrostructures of science, Latour is interested in what scientists actually do. To focus only on the macro level of science has not only developed societal understandings of science's importance, it has also created a mystique and lure that does not allow non-scientists the opportunity to understand what scientists actually do to construct knowledge. Anthropological approaches allow for an increase in "our understanding of the complex activities which constitute the internal workings of scientific activity" and offer opportunities to "penetrate the mystique of science and to provide a reflexive understanding of the detailed activities of working scientists" (Latour and Woolgar 1986, 8–9). This approach to science undermines some traditional and cherished assumptions about science and Western societies. First, Latour's approach to science helps redefine the assumptions of what anthropology is. "We envisaged a research procedure," Latour and Woolgar (1986, 28) announce as they discuss their entrance into the Salk laboratories for a 2 year study, "analogous with that of an intrepid explorer of the Ivory Coast, who, having studied the belief system or material production of 'savage minds' by living with tribesmen,…eventually returns with a body of observations which he can present as a preliminary research report." This statement places the culture of science on the same playing field as other, non-Western cultures anthropologists studied; greatly expanding anthropological notions of culture and chipping away at Western notions of superiority. This begins Latour's journey into the study of the modern that continue to this date as he studies legal and religious cultures in the West. Second, Latour challenges the notion of what science is. Latour and Woolgar in *Laboratory Life* proclaim that what they are doing in the Salk Laboratories is exactly what natural scientists do. In the quote above they offer to spend time in one site to collect data and make observations that can be utilized to construct a research agenda. They offer more parallels to natural science as well. Latour and Woolgar (1986, 30) want to provide reflexive observations that will help not only natural scientists but also social scientists and humanities scholars to understand "that observers of scientific activity are engaged in methods which are essentially similar to those of the practitioners which they study." This is part of the demystifying process Latour and Woolgar (1986, 31) wish to cover in order to demonstrate that "scientific activity is just one social arena in which knowledge is constructed." Acknowledging science as a creative activity places in doubt the traditional notion as to what scientists do at their workbenches. No longer is it valid to accept that scientists are unique in their approaches to discovering universal truths about nature. From Latour and Woolgar's perspective scientists are creators just like an anthropologist or a visual artist. So what is it that scientists do in their laboratory and other cultural settings? This is the task Latour sets out to do in his many works on scientific matters.

As creators or constructors of knowledge, the laboratory process is often a matter of turning chaos into an order that is acceptable not only to the individual scientist and laboratory team but to the wider community of scholars. From an outsider prospective it might appear that a scientist is taking "a body of practices widely regarded…as well organized, logical and coherent," and is really a process that "consists of a disordered array of observations with which scientists struggle to produce order" (Latour and Woolgar 1986, 36). To understand this process of creation, Latour proposes to do as the scientist does. The newcomer to the laboratory, either as an anthropologist like Latour and Woolgar or as a specialized natural scientist, has to enter into the culture of constructing knowledge. "It is clear," for Latour and Woolgar (1986, 37) "that when seen through the eyes of a total new-comer, the daily comings and goings of the laboratory take on an alien quality....In order to make sense of his observations, the observer normally adopts some kind of theme by which he hopes to be able to construct a pattern." The observer creates order, but this is only the beginning. For this order to be considered valid and fac-tual, it has to be accepted by others. "If he can successfully use a theme to convince others of the existence of a pattern, he can be said, at least to have explained his observations." But! "It is not enough simply to fabricate order out of an initially chaotic collection of observations; the observer needs to be able to demonstrate that his fabrication has been done correctly or, in short, that his method is valid" (Latour and Woolgar 1986, 37).

This process of proving one's fabrication of an order through a valid approach as a result of meaningful observations is only the beginning of the whole construction of knowledge process. Now, the scientist has to leave the laboratory and although conditions for proving a "fact" might be difficult to establish in the friendly territory of a laboratory, the outside world can be brutal. Before we leave the laboratory there is still another dimension to cover in the laboratory that is an important contribution by Latour. When scientists construct an orderly system of fact creation, they are never alone. There are, of course, the other scientists, the technicians, the post-doctoral students, the doctoral students, the funding agents, and other guests from anthropologists to science journalists. There are also other important players in the whole process: the non-human actants contributing to the whole creative process of knowledge construction. There are the machines providing readings for the technicians to chart, the equipment to conduct the initial experiments, the animals to take the initial risks, and other non-human objects that serve as conduits to, replacements of, simulations of, and fragmented pieces from "nature" that must be labeled, sorted, interpreted, and presented as "evidence" or "facts" important to the construction of knowledge. As Latour (2004, 35) notes in *Politics of Nature*:

> As soon as we add to dinosaurs their paleontologists, to particles their accelerators, to ecosystems their monitoring instruments...we have already ceased entirely to speak of nature; instead we are speaking of what is produced, constructed, decided, defined, in the learned City whose ecology is almost as complex as that of the world it is coming to know.

As soon as material objects are added into the process of understanding then a whole new world is invented that must be accounted for. If a scientist is interested in establishing a "fact", a key challenge is not only convincing other scientists of the validity of these constructed "facts" but another important point is to figure out how to make these non-human objects communicate when only humans speak? For Latour (2004, 67), scientists "have invented speech prostheses that allow nonhumans to participate in the discussions of humans...The formula is long, to be sure; it is clumsy and turgid; but we find ourselves in a situation where a speech impediment is preferable to an analytical clarity that would splice off mute things from speaking humans in a single stroke." The life of a scientist is sure getting more complicated when the anthropologist is allowed to enter the room. To understand how scientists construct knowledge and create facts, non-human actants have to be taken into account and understanding this process for scientist and non-scientist alike is not easy but it is necessary in order to understand the culture of science.

Peers outside of the laboratory only make this process that much more difficult. They too have non-human actants shaping their knowledge construction processes, anthropologists mucking up their everyday lives, and their own set of "facts" to invent. Sometimes these scientists have to come out of their laboratory to seek broader validation of their creations. When they do scientists enter into yet another phase of knowledge construction. The easiest route to present one's "facts" is to present a paper at a conference. Maybe there will be questions, but usually not, maybe some rivals will be present but probably they do not want to be seen supporting the competition. Sometimes though there are controversies at conferences over a paper represented but this usually only marks a moment for gossip or chit-chat inbetween sessions. The real struggle for establishing "facts" is in the peer review journal process. Here everything matters. What journal will a paper be submitted to, who will review it, what words are chosen, what do the footnotes reveal? Do these footnotes demonstrate enough information to establish that a scientist is naming the right names, proving a valid method, and establishing original work? As Latour (1987, 25) in describing the writing process of knowledge construction, "A sentence may be made more of a fact or more of an artifact depending on how it is inserted into other sentences. By itself a given sentence is neither fact nor fiction; it is made so by others, later on." It matters then who the scientist cites, what lineage one is drawing upon, what alliances one is trying to construct when trying to establish a scientific "fact." How this attempt to assert a "fact" into the scientific lexicon and the "decision about whether it is a fact or a fiction, depends on a sequence of debates later on" (Latour 1987, 27). The debate and struggle over "facts" and sentence word choice demonstrates to Latour (1987, 29) that this process of scientific knowledge construction is a "collective process" that has to include non-human and non-scientific actants as well. There is even ample room for curriculum scholars to join the other actants.

2.3   In Search of Epistemic Things

Latour's anthropological excursions into the scientific laboratory opened wide the mysterious black box where science was done. With Latour the macro level of science was put to the side for a while and the minute details of the micro level attracted the attention of science studies scholars for the next couple of decades. One of those scholars is Hans-Jörg Rheinberger. As deep into the micro level as Latour tries to burrow into, Rheinberger tries to go even further. Like Latour, he is also interested in those moments of knowledge construction that happen before "facts" are established as facts. It is here that science happens and it is this realm that Rheinberger calls the epistemic thing.

In his influential two part articles on "Experiment, Difference, Writing" published in *Studies in History and Philosophy of Science* in 1992, Rheinberger borrows Derrida from philosophy to explain how science is done at the everyday level in order to explain the numerous steps and details needed to produce facts. Rheinberger begins with experimental systems to build his model of fact construction. Rheinberger (1992a, 309) considers "an experimental system as the smallest functioning unit of research, designed to give answers to questions which we are not yet able clearly to ask. It is a 'machine for making the future.' It is not only a device that generates answers; at the same time…it shapes the questions that are going to be answered….It cogenerates…the phenomena or material entities and the concepts they embody." As is the case with Latour, again the scientific process is not one of clarity, certainty, and establishment of bedrock facts. Science is presented more as a groping; a struggle to find anything that can give an investigator a footing to stand and then build from there. An experimental system is, following Latour, a reflexive entity that not only gives answers to problems but also shapes those problems simply because this is a system. As a result Rheinberger's experimental system is very similar to autopoiesis, feedback loops, self-reflexity, perspectivalism, and deconstruction. It is a system built upon an uncertain foundation, an ecology open to outside influences but shaped by its own culture and traditions. When experimental systems are studied "two different yet inseparable structures or components can be discerned. The first can be called the *scientific object* under investigation, or the 'epistemic' thing.' The second can be referred as to the technological identity conditions, or the *technological object(s)*" (Rheinberger 1992a, 310). The scientific object is a "physical structure, a chemical reaction, a biological function whose elucidation is at the center of the investigative effort"…that is "in the process of being materially defined." The technological object, on the other hand, is "characteristically determined. They perform…according to known regularities. They contain the scientific object in the double sense of the word: they embed it and they restrict it" (Rheinberger 1992a, 310). Breaking from a Kuhnian assumption that it is theory that drives scientific progress and defines scientific problems, Rheinberger is asserting that in part it is the technology or the parameters of the experiment and all the material objects involved in an experiment and its later system of ideas that shapes theory and the problem. It is the technological object that controls the epistemic thing and creates the conditions under which an epistemic thing can eventually be labeled as a "fact." As a result how the technological objects are framed, established, created, and constructed will shape the meaning of the epistemic thing. What Rheinberger presents here is a detailed account of how scientists move from initial construct of a problem, through the whole process of inventing a protocol for experimentation, to eventually the establishing of a scientific "fact." Yet, the explanation of the process of scientific work has only begun.

Although epistemic things and technological objects are considered to be interchangeable at times in the process of establishing a fact, they both serve different purposes. "Scientific activity is scientific only and just in that it aims at producing future, whereas technological constructions aim at assuring present… technological construction rests on identity in performance; scientific construction rests on difference…A technological product basically answers the question which is implemented in its construction….In contrast, a scientific object basically is a question-generating machine" (Rheinberger 1992a, 311–312). As the epistemic thing and technological objects interact they enter into a very symbiotic relationship in which in "the process of making science, new possibilities of technological construction arise. Knowledge in the form of technological objects enters the social process of reproduction. They function as tools for production or serve as items of consumption. But they must also re-enter the research process. In this way technical tools define the 'system' of investigation…they become part of the controlled boundary conditions of the experimental system" (Rheinberger 1992a, 312).

Once the epistemic object is targeted and the technological parameters are established thereby defining the meaning of the epistemic object and its limitations, the experiments are conducted, the questions and the data are generated. It becomes, as Rheinberger calls it, a "tracing game."

> This is a question of representation, or of translation. A scientific object investigated by an experimental system is deployed and articulated within a space of material representation… The structure of the scientific object contained and contended about in the experimental setting constitutes a model. The model is a structure through which the noise produced by the research arrangement is translated into voice, trace, and writing….Representation, then, is equivalent to bringing a scientific object into existence. Once realized, the model is not interesting per se; it remains interesting only as a tool, as a technical device for constructing novel research arrangements. (Rheinberger 1992b, 390–391)

It is through the process of representation that the epistemic thing is given foundation and meaning, and in order to understand how science is done, one needs to understand how the model acts simultaneously as the technological parameters that shapes and limits the questions that can be asked and defines how they are asked, and, at the same time, provides the parameters to construct meaning of the epistemic thing as it moves to become an established, accepted fact. At the same time while a model is providing the parameters of what questions can be generated, it is also providing the data. But any model is limited in its power because the success of any model is not in its comparison to nature, what is out there in "reality". As Rheinberger (1992b, 392) notes "Nature as such is not a reference point for the experiment; it is even a danger….Consequently, the reference point of any controlled system can be nothing else but another controlled system. The reference point of a model can be nothing else but another model." What the model does is provide marks, call them data, facts, outcomes, points, whatever you like, "through measuring devices and technical arrangements" that open "a space of representation" (Rheinberger 1992b, 393). It is here that science gets written. Yet there are different levels in which it is written.

What is usually focused on when the issue of writing science is broached is the final product presented in journals and eventually in textbooks establishing the state of a scientific field. Yet as Rheinberger points out there are many authors in the writing of science that are important to understand. The first was already mentioned, the passive voice writer of research. When scientists are writing in the passive voice they are not writing, it is part of the Modest Witness persona to be discussed later when Donna Haraway is mentioned. As Rheinberger (2003, 312) notes "when scientists write research articles, they are as primary mediators; that is they pretend to let their objects speak to the small community of those working on similar matters." In research articles epistemic things are cemented as facts, actants are established as parts of nature, paradigms are confirmed or challenged, and research agendas are established or confirmed. It is through passive writing that authority is established, the authority to define nature and facts or, to put it another way the authority to pretend that a scientist has no authority except the gift of letting nature speak through them.

Active voice is a second writing mode. This is when a scientist is allowed to be a human, contributing to a human field of knowledge through a human creation called writing. Active voice writing is not as important or as authoritative because it does not hide the humanness of science and expertise is not established since almost every human can write, even poets and fiction writers. Yet in order to be able to write in the active voice one needs to have authority within a field of knowledge. As Rheinberger notes autobiographies of scientists are written in the active voice and very rarely does one find in the bookstore or on Amazon an autobiography of a laboratory technician or a doctoral student. Autobiographies are written by those scientists who have changed the field, earned a high status, won the important prizes, published in the premier journals, and held the prestigious positions.

The third writer voice found in science is important for Rheinberger and it is the writing voice found in scribbles. Scribbles are constructed at the bench as laboratory work is happening and is the first writing that takes place when science is being created. "These traces reach from jotting down ideas to drawing sketches of experiments, recording data, arranging data, processing data, interpreting experimental results, trying out calculations and designing instrumentation" (Rheinberger 2003, 314). Scribbles are a free flow of ideas, a free write exercise in the sense that there are rarely restrictions on what can be written down and this writing is happening as epistemic things are appearing or either disappearing as an unproductive research path.

A fourth authorial voice for Rheinberger comes in the form of standardization and the general culture of the laboratory. Standardization comes in the form of set protocols established through manuals that inform everyone in the laboratory how samples should be taken, data recorded, species handled, and results categorized. These writings "preserve for generations of experimenters what has proven to be successful. They are the forms of life into which newcomers are socialized" (Rheinberger 2003, 318). What all of these forms of authorial voices demonstrate for Rheinberger are the moments of authorification in which the authority for a scientist to write is established even before the scientist becomes the passive mediator of an established fact.

Rheinberger's work serves as a call for curriculum studies to enter into science debates. His focus on models reminds us of their importance in shaping scientific knowledge, especially in economics and climate science. Ignoring the role that models play, as climate change deniers demonstrate, comes at our own peril. It is in these models and how they are built over a long period of time determine the data that is mined from these models and the conclusions that are drawn to inform policy. While climate change deniers accept the complexity of economic models and the assumptions that are built into them often at the expense of everyone's well-being, they do not seem to wish to take the time to understand how climate change models are constructed and the importance of their data. Will curriculum scholars? Rheinberger's focus on writing is also an opportunity to explore how scientists navigate through an important question for curriculum scholars: What knowledge is worth most?

2.4 Institutionalizing Science

Thus far, the discussion has been geared towards the individual scientists and perhaps, as in the case of Latour and Rheinberger, their relationship with other scientists and non-human actants. Yet, almost every scientist in a field is institutionalized either in an academic setting, a research institute such as the Max Planck Institutes, an industrial research unit, or a government agency. Virtually gone are the days when most science was completed outside an institution, yet even then institutions played a significant role in guiding and defining science. For instance Robert Boyle was able to conduct much of his work on his own research without any institutional encumbrances, but when it came time to prove the importance of his air pump experiments he was very dependent on the Royal Academy of Science in order to control the meaning of his experiments and the establishment of priority. The same is true for Newton and Darwin. Both were able to do much of their thinking outside of institutional settings but when it came time to establish priority Newton was dependent on the Royal Academy of Science to establish that he not Leibnitz was the inventor of Calculus, and for Darwin he was dependent on the agreements of British scientific organizations to establish that Darwin created the theory of evolution while Alfred Wallace made substantial contributions. Institutions always played a role in scientific knowledge construction, and Timothy Lenoir is an important scholar studying the connections between science and institutions. His work is an attempt to understand the macro and micro levels of science.

All the scholars mentioned thus far recognize that there are cultures within science that shape how science is done. Understanding these cultures is essential in order to do science and navigate through universities, laboratories, industries, and government bodies, Lenoir is no different except he places more emphasis on the institutions to determine scientific cultures. "Institutions guide, enable, and constrain," according to Lenoir (1997, 1), "nearly every aspect of our lives. Within scientific fields, professional life takes place entirely within a context nested, overlapping, interacting, sometimes conflicting institutions." Institutions are the sites where scientists not only do science, but also where science is defined by and for them. Yet most science studies scholars have ignored the impact of institutions on scientific work. Traditionally scholars such as Robert Merton and Joseph

Ben-David focused on macro level structures of science creating what can now be called a classical notion of science. Merton and Ben-David constructed science as "a rational inquiry into nature in terms of logical inference aimed at finding universal laws, precisely written in the language of mathematics, and the prediction of new empirical facts deducible from theory confirmed by observation and experiment" (Lenoir 1997, 4). This tight definition of science allowed early sociologists to define their research agenda from other realms of possible sociological study such as politics, religion, economics, and culture. This lead Merton and Ben-David to assume that science, as an entity free from these other realms of life, was defined by "realism, objectivity, disinterestedness, and autonomy" (Lenoir 1997, 5). These four principles allowed early Sociologists to create the mythical lure that scientists were merely trying to understand reality free from any human taint and held no interest in shaping politics in any way thereby allowing them the freedom to conduct science as their professional expertise guided them to do. Any attempt to interject politics, religion, economics, or even culture into the scientific process "diverts science from its goal of producing truth" (Lenoir 1997, 5). Although he wishes to continue the tradition of looking at the macro level of science, Lenoir views this tradition of the sociology of science as problematic. Lenoir is interested in how science is not isolated from other realms of society but specifically implicated. As Lenoir (1997, 3) notes "I emphasize the manner in which science as cultural practice is imbricated in a seamless web with other forms of social, political, even aesthetic practices, and I treat the formation of discipline and scientific institutions as sites for constructing and sustaining forms of social and cultural identity." With this goal established Lenoir combines the macro (science, politics, culture, and economic) with the micro (he specifically studies the influence of science on art and vice versa in nineteenth century Germany in building both scientific and artistic institutions, and he focuses on German scientists who developed institutes of scientific research that later became the model for Kaiser Wilhelm Institutes) in order to demonstrate how science emerged as an essential component of modern life. In order to accomplish this Lenoir (1997, 9) offers "an alternative to theory-dominated history of science— by forcing us to consider the historically situated, time-dependent character of plans and actions. The focus on practice shifts our gaze to the mundane: to the construction of instruments, the manipulations of experimental apparatus in the time and space of the laboratory."

Andrew Pickering is a sociologist of science who has adopted and adapted Lenoir's approach and by focusing on the nineteenth century dye industries demonstrates how science is institutionalized and unquestionably bound and intertwined with other dimensions of society. Instead of focusing his critique on the limits of Merton and Ben-David's sociology of science, as Lenoir does, Pickering focuses on the limits of traditional Durkeimian sociology that viewed science and technology as "presumably by virtue of their rationality and materiality, other than the social, beyond the purview of the sociological gaze" (Pickering 2005, 355). In place of a traditional Durkheimian approach to science, which is to say the approach of not approaching science as a legitimate site for sociological study, Pickering suggests creating what he calls a double centering. Double centering suggests "on the one hand, we have a history of science and technology evolving autonomously from the social, the centered, say on the rationality of science and the powers of technology…On the other hand,…one usually finds a story of social accommodation to technology, where sociologists can continue to operate their usual social centering, but now with a moving boundary condition" (Pickering 2005, 356). This approach allows Pickering to acknowledge that science and technology are separate entities from other sociological categories such as the economy, politics, religion, and culture, yet recognize that all these social phenomenon interact, thereby making them all legitimate sites for sociological investigations, with each other and a result reshaping each other. Pickering (2005, 359) uses the terms translation and tuning to demonstrate how science interacts with other sociological entities. Translation "refers to the movement of elements from one setting [science to university or university to industry] to another in the space of multiplicity; tuning refers to the open-ended adjustments that typically prove necessary if translations are to be successful." In order to contextualize his theory of science, Pickering focuses on three institutions (the Dye industry, the university, and the state) to demonstrate how all three are independent of each other yet shaped by the others over the long term.

Pickering focuses on the European dye industry between the 1820s and 1860s not only because it went from creating dyes out of organic material to working with inorganic coal residues to create synthetic materials but also because "The social world was thus changed and rearranged in Perkin's [reference to the British industrialist William Henry] industrialization of his achievement…At the same time, it was clearly not purely a social event. It hinged upon new flows and transformations of matter" (2005, 366). Through synthetic dyes and the industrial organization of their production, new social relationship were constructed between owners and workers but more importantly between industrialists, chemists, universities, and new material objects. The dye industry created a "scientific-industrial assemblage…" that is what Pickering calls a "double translation. Materials, such as dye samples and money (as fees to expert witnesses) flowed from industry to the laboratory and information, analyses and new dyestuffs and opinions on syntheses, flowed back…Science and industry not only came together, then; they were reciprocally transformed in the patent suits over new dyes…The becoming of each helped structure the becoming of the other" (2005, 372). Eventually the universities became involved. Initially, industry simply hired the chemists trained at the university and then trained them to work in the industrial laboratories. The training was both specific to the chemistry needed to produce and discover new dyes quickly and to indoctrinating new chemists to the culture of industry. "New recruits," Pickering (2005, 394) notes, "were processed through a lengthy in-house training program that began with work on printing and dyeing, moved through known syntheses to analyses and syntheses of rival products, and slowly and gradually in its second year, passed into genuine and specialized research in some branch of color chemistry." By the 1880s, these arrangements changed when the dye industry became dissatisfied with chemistry education in universities. As a result, the German Chemistry industry began "to sponsor reform at universities and Hochschulen" (technical high schools) (Pickering 2005, 394). These relationships between industry, scientists, universities and the state demonstrate for Pickering that industry pushed changes that took place in chemistry and the universities. "Chemical theory was not prior to material practice. Rather it functioned in a retrospective/prospective manner, summing up existing achievements as a fallible and revisable guide to future research. The criterion of a good theory was that it worked out in material practice" (Pickering 2005, 387). For Pickering this demonstrates the limits of Durkheimian sociologists who excluded science from their areas of interest in understanding modern society. Studying the rise of the chemistry industry in England and Germany demonstrates that sociologists cannot understand major changes in modern European society in such areas as worker's lives, the rise of the professional class, university curriculum reform, or state policy formation without understanding the role science and its material objects played in shaping changes in modern society. I would add to Pickering's thesis that curriculum scholars cannot understand contemporary public school deforms pushed by corporations without knowing the history of science and university reforms.

2.5 Comparative Literature Comes to Science Studies

As is the case with the other scholars I discuss in this chapter, the individuals adding to my understanding of science in society are not necessarily the first scholars within their fields to enter into the realm of science studies. They are, however, some of the most influential in shaping the field of science studies. N. Katherine Hayles is a prime example. Comparative literature, like many disciplines such as geography, economics, history, information technology, chemistry, is born from a crisis or an epistemic shift in knowledge. Anyone who is aware of the history of the university understands that as an institution it is dramatically changing, and anyone who has read Bill Readings' (1996) *The University in Ruins* knows literature is in a crisis. Literature, prior to the current shift from nation-states defining identity and dictating policy to multinational organizations, mostly corporations, shaping identity, was valued as an important avenue to shape the meaning of citizenship in a nation-state. This citizenship is no longer valued in defining individuals, literature has lost its value and purpose in society and the university. In order to reinvent itself, literary scholars have redefined what they do. Some of these scholars have expanded the meaning of comparative literature. Traditionally, comparative literature meant comparing pieces of literature from genres, time period, or perspectives. Today comparative literature is defined, at least in part, as a study of Literature in relationship to other fields of knowledge. J. Hillis Miller and Peggy Kamuf have developed their careers by joining literature with Derridean philosophy, Andrew Ross left an early career in the study of poetry and now devotes his scholarship to literature and american studies, and Mark Hansen has developed a career comparing literature to information technology and new media. Katherine Hayles has lead the way making connections between literature and science. As a result, these scholars have redefined literature and demonstrated the discipline's essential role in understanding other fields of knowledge.

An essential concept to understand Hayles contribution to Science Studies is the Cosmic Web. It is not only the title of her first book, *The Cosmic Web: Scientific Field Models & Literary Strategies in the Twentieth Century*, but also a key concept guiding her early work. For Hayles there is a unique relationship between scholars in distinct and often isolated fields of knowledge. In her works, Hayles has noticed that scholars within Literature and the Sciences often grapple with similar ideas during the same time period even though these scholars do not know the other fields are studying a similar phenomenon. This she has found to be the case in fiction, field theory, chaos theory, cybernetics, information technology, neuroscience, and digital media. A cosmic web, in Hayles (1984, 15) own words is a "dance, a network, a field—the phrases imply a reality that has no detachable parts, indeed no enduring, unchanging parts at all. Composed not of particles but of 'events,' it is in constant motion, rendered dynamic by interactions that are simultaneously affecting each other." Part of this constant motion is for fields of knowledge to move in their own trajectory, shaping their own field and covering topics, objects, and projects that other fields are also covering but in their own way. One of Hayles' first attempts to demonstrate how literature and the sciences dance in the same web was to look at literature and chaos theory. In *Chaos Bound*, Hayles (1990, 3) writes "that certain areas within culture form what might be called an archipelago of chaos. The connecting theme is a shift in the way chaos is seen...not as an absence or void but as a positive force in its own right....Concerned with the physical science as well as literature, the study investigates language's power to constitute reality, and reality's power to constrain and direct language." It is an attempt to demonstrate how science shapes language and language shapes science.

The recent "turn" in Hayles' work since 1999 has been the posthuman and digital technology. In her work *How We Became Posthuman*, Hayles challenges the humanist and modernist notion that favors the mind over the body and separates the two. Her goal is to reconnect the two and demonstrate the materiality of the mind. In order to chart the disappearance of the body from the development of humanism and scientific movements, Hayles looks at the evolution of cybernetics as a field of knowledge that devalued the body. "I see the deconstruction of the liberal humanist subject," Hayles (1999, 5) writes, "as an opportunity to put back into the picture the flesh that continues to be erased in contemporary discussions about cybernetic subjects." Yet, she constructs her argument in a manner that does not devalue the sciences or construct a ghoulish scenario in which information technologies overtake humans. As with her other work, what Hayles demonstrates is how the natural sciences use language to construct a reality that often is nothing more than a complicated model based in conjecture and assumptions that shape reality but have nothing to do with any actual reality. This is certainly the case with the humanist subject and the disembodied human in cybernetics. Worlds have been constructed out of and because of humanism and cybernetics, thereby constructing a reality that with the help of power became established "truths" and "ways of knowing," but as soon as these worlds were challenged and their surface scratched nothing firm appeared underneath but human assumptions, beliefs, and desires. The same can be said about neoclassical economics. Many models and theories have emerged from a neoclassical mode of thinking, but nothing can hold universally; certainly not as universally as neoclassical economics insist. Instead what one finds in neoclassical economics is creative ways to construct and justify economics through model creation, but nothing that can in a satisfactory way explain major aspects of economic life let alone other dimensions. What is exposed is a lot of wishful thinking, ideological assumptions, and political power plays. This by no means suggests economics is smoke and mirrors with no foundation or real purpose as a field of knowledge. Neoclassical economics explains plenty, economists just too often step beyond their bounds and try to explain more than they can, explain away what their models miss, and dismiss alternative worldviews that challenge neoclassical approaches. Hayles demonstrates how this happens in most scientific paradigms. In regards to the rise of cybernetics after World War II, Hayles points out how the path scientists selected was not the only one they could have chosen. Had the scientists selected an alternative path the outcomes would have been different in understanding the nature of information and what it means to be human.

2.6   The Return of the Macro-Level

A new approach to science studies has emerged in the late 1990s with the return of macro level topics that impact the making of science as it relates to society. Unlike early macro level sociologists like Merton and Ben-David there is no attempt to romanticize science as objective, neutral, and apolitical nor like Kuhn to protect science from outside forces such as lay people and policy makers. Instead, what scholars such as Dominique Pestre and Helga Nowotny try to do is demonstrate how politics and policy formation as well as academic politics shapes what science does and how science works. At the same time these scholars do not abandon the micro level of science but do offer a "soft" critique of Latour and other scholars who created the actor network theory. In this Chapter I will focus primarily on Pestre and introduce Nowotny's work in chapter five because she offers important thoughts on the (de)coupling of science and democracy.

There is an old adage in physics that says if one does not make a contribution to the field by the age of 30, then one never will. There are plenty of examples of physicists contributing major theories and understandings while in their 20s, but I have never heard of anyone writing about physicists who failed to contribute to the field and then move on to something else. What has interested me with those physicists later in their career after that watershed mark for creativity and innovation has passed, is what do they do? Some as we will see in chapter five like the "trio" of climate change deniers join a conservative think tank and spread ignorance and doubt. The "trio," however, are not the norm. What I have found is some of these physicists leave the field and often turn to science studies. There is the rare exception like Peter Galison, whose work constitutes a major contribution to the history of science, who set out to become a historian of science but also earned a Ph.D. in physics. Many "retired" scientists start a second career in the sciences and begin to report on scientific developments for a wider audience. Freeman Dyson (1997, 1999) has covered recent scientific and technological developments and explaining their importance to society. R.C. Lewontin (2000) has written on the impact of biological determinism and progress in genetics as they impact society, the mathematician John Casti (1998, 2003) has written fictional encounters between twentieth century scientists, philosophers, and mathematicians to discuss such as scientific speculation and the limits of knowledge, and the biologist Evelyn Fox Keller (1983, 1995) has written on underappreciated female scientists and the role of language in shaping science. Pestre emerges from the sciences in this same vein of writers and scientists. A practicing physicist, Pestre now contributes major works on the implications of science studies on politics and policy in Europe and beyond.

Prior to Pestre, Nowotny, and a new generation of science studies scholars, there were two traditions that offered critiques of the traditional approach to science represented in Merton and Ben-David's work as well as previously unmentioned work such as Karl Popper's, Ernst Mach's, and the Logical Positivists. The first tradition lead by sociology of scientific knowledge scholars Harry Collins and Barry Bloor, "wanted…to describe science 'as it is really done'…but they also wanted to act politically—denouncing science as an institution while revealing its 'true' nature" (Pestre 2004, 352). It was an attempt to expose the ideological underpinnings of science that worked from the assumption that science has a true nature and it was time for it to be exposed since the scientists and those scholars who studied science were not interested in doing so. The second tradition was already mentioned and is embodied in the work of Bruno Latour, Michel Callon and other anthropologists. As with the first tradition, Pestre acknowledges their major contributions to the field of science studies but also discusses their weaknesses. "Their aim," according to Pestre (2004, 353), "was to contest the traditional notion of 'facts' or 'proofs,' and to study how these were 'negotiated' within the laboratory and within scientific communities." The problem with this approach is it too easily accepted the commodification of science that neoliberals were imposing on the science fields. In Latour's approach there is no critique or "calling out" of neoliberal's attempt to turn science into another commodity that can be exchanged like it were money, automobiles, or a piece of pizza. One reason for this acceptance of a neoliberal definition of science for Pestre (2004, 359) is that the supporters of micro level approach to science "tend to favor [a]'charitable' reading of everyone".... "Since the aim is to be understanding...no judgment or comment about what is invoked by actors is 'naturally' forthcoming." Pestre's critique of micro level requires science studies scholars to self-reflectively answer how their work privileges certain ways of knowing and what impact these privilegings have on establishing policy and setting up notions of "truth" and "knowledge."

In replace of a macro or micro level approach Pestre constructs a multi leveled approach that explores the ways in which science is influenced by politics, science can influence politics, and how science impacts society. Part of the neoliberal approach to science is to privatize knowledge. The dominance of this perspective can be seen in the rise of patents to protect not only products created in a laboratory or by an individual scientists but also to protect tools used in certain techniques and new organisms or pharmaceuticals created in the laboratory. The privatization of knowledge is also marked by the creation of transfer of technology agreements and technology control offices that are called to monitor the transfer of ideas. What Pestre tries to do in his work is offer alternatives to this way of thinking in order to guarantee that science will be as responsive to democratic needs as possible. As Pestre (2004, 365) notes, "if we accept that there is a growing privatization of knowledge today, then working to guarantee the plurality of knowledge-production institutions as well as protecting their different moral economies and value systems might be central to any political action." Pestre challenges a traditional approach such as Kant's in *Conflict of the Faculties*, Weber's (Eisenstadt 1968) "Science as a Vocation" or Kuhn's *The Structure of Scientific Revolutions* in which science is viewed as neutral and apolitical. For Pestre, science should be political in helping all voices be heard in science policy and in areas where science can help improve the well-being of individuals such as when dealing with clean water or sustainable environments. This marks a major shift from any previous science studies approaches mentioned above and joins feminist and post-colonial readings of science that will be discussed later.

If knowledge is becoming more privatized then a new approach to knowledge production is important. Drawing from the work of Helga Nowotny and her colleagues, Pestre suggests there are two modes of knowledge; fittingly mode 1 and mode 2. Mode 1 is the classical approach to knowledge that separates pure science from practical science, thereby protecting the scientist from politics. In mode 1 the scientist is disinterested in who gains financially from their inventions and discoveries as long as the proper scientist is granted priority. The scientists in mode 1 are motivated by curiosity and a strong sense of community to share new knowledge. In Mode 1 (see Nowotny et al. 2003, 179) there is a "hegemony of [the] theoretical or, at any rate, experimental science; by an internally-driven taxonomy of disciplines; and by the autonomy of scientists and their host institutions, the universities." Mode 2 is a revision of the first "model" and represents a more nuanced approach to understanding and challenging the rise of privatized knowledge production. Pestre (2003, 247) acknowledges that political, economic, and societal powers have always been interested in knowledge production including "gadgets and material techniques—arms, objects to be sold, tools to improve production; it has been of interest to planning and management of military operations, social action, political control, and financial awards." Knowledge and the scientists who construct it have always been political or influenced by political powers that shape scientific agendas and the funding of research. The issue for a mode 2 approach to knowledge production is how can more people influence the power structure and thereby shape science policy? It is by no means a given that since post-industrial nations are shifting more to what is now being called knowledge based economies that this "does not automatically lead to an open system of exchange and cooperation" when the dominating logic or rationale behind a knowledge economy is based on "the control of markets and profits" (Pestre 2003, 254). In a knowledge based economy the individual is constructed as a consumer and as a result their roles in science policy and political decision making are foundationally limited. The most glaring examples of the individual as consumer are in food production and pharmaceutical drugs. In both cases citizens in a so-called democracy are not involved in the decisions as to what drugs will be created or subsidized nor what diseases will be studied. Instead individual citizens become what Kaushik Rajan (2006) "patients-in-waiting" or test sites in which a new drug can be administered and studied. The same applies to food production. Individual citizens who wish to know the content of their food are resisted by corporations who have turned food production into a mega business whose primary goal is not healthy individuals but profit and market control. What has emerged is what, following Ulrich Beck, Pestre calls "the risk society." In a risk society, individuals assume all the risk while corporations claim the profits. If the overproduction of food like substances dominates one's local grocery store and that is all one can afford because they are heavily subsidized by corporate controlled governments, then, as the neoliberal ideology insists, individuals, as the rational beings, are the only ones' responsible. If by chance an individual citizen wishes to petition the government and seek policy changes to protect them from predatory corporations, the neoliberal ideology has a response too. In order to control how local, federal, and international governments respond to individual needs, when they conflict with corporate demands, is not only to directly control the government, but also to starve the government of the necessary funds and authority to control corporate actions. Corporations also use the willful propagation of ignorance to control citizen actions as well. This approach can be found in the Tobacco industry and in the industries with a vested interest in denying climate change.

In this battle between corporate demands and individual citizens, Pestre believes the role of scientific expertise is important. How shall this expertise be used? Pestre (2003, 256) reminds us that "expertise is not a neutral political entity" and in any political struggle over scientific issues corporations will not only have their majority share of legal authority but they will also have their more than fair share of expertise representing any position they wish to hold. Pestre offers a few suggestions as to counteract this unbalanced power relationship. It is important for citizens to reclaim the regulatory agencies charged with the assurance of food and drug safety as well as environmental safety. Corporations have infiltrated the ranks of regulators by having their leaders serve as government regulators. The neoliberal ideology has also gutted the funding for independent government and university research on the safety of new drugs and newly invented food like substances. This means the corporations will do their own research; hiding any damaging results and highlighting and marketing any positive outcomes. Another suggestion Pestre makes is the importance of scientists and policy makers to consult and work with citizen boards who play a role in constructing science and government policies. There is evidence that corporations have also corrupted this idea as seen by the pharmaceutical industry creating fake parent organizations to push their drugs. The imperative is clear for Pestre as to the connections between science and a vibrant democracy. "The parallel idea that 'civil society,'" Pestre (2007, 418) contends, "should play a central or renewed role in the management of techno-science in/and society is…an interesting normative posture and a heuristically illuminating notion….If the idea is to favour the active presence of 'ordinary citizens' in the management of the polis,… then the notion is certainly central." If we wish to continue to foster democratic ideals and the growth of an intellectually vibrant society then these suggestions Pestre offers should not be taken as mere polite suggestions from another academic, but rather as necessary steps in order to finally take policy making and knowledge creation back from the oligarchical threats (both from the plutocratic dictators and corporate tyrants) that so-called democratic nations face at this very moment.

2.7   Of Perspectivalism, Modest Witnesses, and Strong Objectivity

There are certain scholars who have had a major influence on how science is viewed but they cannot be organized around macro or micro, mode 1 or mode 2, or theory versus practice. This loosely fitted grouping would include Nietzsche, Feminist scholars, and Post-Colonial thinkers. All three have shaped my thinking about science and in later chapters you will see how they have shaped my thinking. First, Nietzsche.

When the topic of discussion is Nietzsche and science an important concept is perspectivalism. Perspectivalism is simultaneous acceptance that life, in order to be truly life, is an interpretation. Life is the invention of meaning where there is none and the struggle for life is the struggle to enact one's notion of truth and life in the world. Perspectivalism is the acceptance of the best available interpretation of any aspect of life, in the name of real life, and this acceptance only means the best available for now. In the future the circumstances as to what is the best interpretation will change and one must be ready to jettison old interpretations for better ones. However, just because life is an interpretation and the best interpretations should be accepted by all who possess sound and rationale minds, this does not mean that relativism is what Nietzsche is advocating. Relativism is an impoverished way to live life. All possible interpretations are not equal, and most interpretations ought to be rejected in the name of life. To settle for anything less than the best possible interpretation is to accept a herd or slave mentality when all humans have the capacity for the Will to Power. Babette Babich (1994, 5), explains it this way: "With perspectivalism, Nietzsche offers knowledge an infinite domain, but such a perspectivalism offers knowledge seekers no such infinite and no sure method and no truth." Perspectivalism provides no firm foundation from which to build a worldview, you are alone with your rhetorical skills, marshaling of power and "facts," and interpretive skills to invent reality. The possibilities from this foundational-less perspective is infinite, the meaning of the universe, earth, humankind, and life in general is up for grabs and meaningless until someone or something gives life meaning. Science is just one of the many creative ways that humans have invented to create meaning. The problem for Nietzsche is that the scientists and philosophers of science have ruined science. Science has found a way to deny it is founded on interpretations, banned the infinite from its boundaries, and limited itself only to that which can be tightly controlled and manipulated so some universal law, which is never universal, can be proclaimed.

Why has science disguised its perspectivalism as truth and why did Nietzsche spend so much time criticizing scientists? It is because they stopped doing science and instead did repetitive, mechanical work and labeled it science. As a result calculation, ordering, classifying, and data collection became the norm for scientists and for Nietzsche this sapped the life out of science. The evidence to support Nietzsche is abundant both during his lifetime and after. In the Nineteenth Century scientists scoured the earth looking for anything to collect. The best, like Darwin and Humboldt, were discrete and had an idea as to what they were looking for, the majority just collected and then brought what they collected in order to count and calculate as if a collection taught them anything about a region, a people, or a history. The only thing it revealed is collectors like to draw inferences from objects, and some cases people, and then deny that any interpretations were being invented in the first place. That is an illness. Science for someone like Louis Agassiz, a great collector of objects for Harvard University, became a pretext to justify racial inequalities and began the strong push towards what we now refer to as scientific racism. When Statistical approaches to almost every state function and many scholarly fields were emerging in the late Nineteenth Century and early twentieth Century, it became a means to testing assumptions and a foundation for making better policy decisions. For some, however, according to Gigerenzer et al. (1989, 107–108), "a strikingly narrow understanding of statistical justification made a significant result seem to be the ultimate purpose of research and non-significance the sign of a badly conducted experiment." This is mechanistic not scientific.

There were other reasons why Nietzsche criticized science as it was conducted in the Nineteenth Century. Science, since at least John Stuart Mill, has succumbed to the sickness of utility. "What?" Nietzsche (1974, 85) proclaims, "The aim of science should be to give men as much pleasure and little displeasure as possible?" This is an utilitarian view of science and of life. To adopt such a view is to forget and erase the reality that "the faith in science, which after all exists undeniably, cannot owe its origin to such a calculus of utility; it must have originated in spite of the fact that the disutility and dangerousness of 'the will to truth,' of 'truth at any price' is proved to it constantly'" (Nietzsche 1974, 281). Utility denies that science is a human creation, filled with, hopefully, as much human passion, will, imagination, and emotion as possible. Yet these human traits are also denied in science. "In science convictions have no rights of citizenship…Only when they decide to descend to the modesty of hypotheses…of a regulative fiction, they may be granted admission…But does this not mean,…that a conviction may obtain admission to science only when it ceases to be a conviction? Would it not be the first step in the discipline of the scientific spirit that one would not permit oneself any more convictions"(Nietzsche 1974, 280). What then is science if it is not about hypothesis testing, utilitarian convictions, or ordering the universe? Like any aspects about human lives, it is about error.

It is an error to think science can create universal laws, collect, calculate, sort, and order the infinite, and create "regulative fictions" and then deny they are regulative fictions. These are only a few errors that make science a human science. "Man has been educated by his errors"(Nietzsche 1974, 174). Some may believe this is why science is such a great educational tool and part of primary, secondary, and tertiary education curriculum. To learn from the errors of the greatest minds is often the best way to learn. This, however, is not what Nietzsche means by error. The idea of learning about science through the mistakes of "great minds" is just another form of hagiography and a grave error. In chapter nine I will develop Nietzsche's aphorism on error, but for now and briefly he means four things when he uses the word error. Nietzsche believes modern science is built on erroneous assumptions including the erroneous assumption that scientists hold no assumptions. Scientists are also erroneous for believing in the idea of the modest witness or that nature spoke through scientists and the scientist did not speak, adopted anthropocentric notions that placed humans above all other animals, and placed all powers in the ideal of the universal when all ideas are contingent. To admit that science is misconceived by these errors is not to admit that science is something unworthy of its name, it is to accept science's importance as a human endeavor, an art, an act in the name of life. Removing these specific errors from science would make it more human and less mythical, but to assume that it is possible to remove all errors only makes science too mechanical and ill. To make science more mechanical and ill is to open science up to the forces of inhumanity in which science becomes part of a death machine used against those people who are vulnerable. So much of what Nietzsche proclaims has become a warning for what science could do in the name of dictatorships, capitalism, and war. When errors are banished and human convictions denied roles in science, science becomes dangerous and a crime against humanity.

Instead of purifying and classifying, science needs to reclaim its artistic roots. As Babich (1994, 206) sees it "for Nietzsche, science is to be conceived aesthetically. While science is not articulated as an artistic or deliberate aesthetic creation it is nevertheless effective...on an artistic level." As artists, scientists reveal their creations and as a result become truthsayers driven by the will to truth. "We," Nietzsche (1974, 266) wants us to believe, "however, want to become those we are—human beings who are new, unique, incomparable, who give themselves laws, who create themselves. To that end we must become the best learners and discoverers of everything that is lawful and necessary in the world: we must become physicists in order to be able to be creators in this sense—while hitherto all valuations and ideals have been based on ignorance of physics or were constructed so as to contradict it." We all need to become the consummate learners so to discover everything and to proclaim our love of life—Amor Fati, the love of fate. To narrowly define the arts as the humanities and the fine arts is not to love life as it is equally true that to define science in a utilitarian and mechanistic manner is not to love life. One either loves all of life or none. "To this day you have a choice," as Nietzsche (1974, 86) declares, "either as little displeasure as possible...or as much displeasure as possible...If you decide for the former and desire to diminish and lower the level of human pain, you also have to diminish and lower the level of their capacity for joy. Actually, science can promote either goal." I stand with Nietzsche and I want science to promote the risks that come with life because this is where true learners and discoverers live fully. Unlike Nietzsche, however, I do think we can live a full life without risking the basics and essentials of life such as medical care, a quality education, and shelter. Science, not as a foundation for more efficiency and productivity, but as an art for creation and discovery is as important as any discipline to richly live a life.

As important as Nietzsche is in my thinking about what science is, feminist scholars of science are just as important. The first "feminist" piece of scholarship I read was Sharon Traweek's chapter "Border Crossings: Narrative Strategies in Science Studies and among physicists in Tsukuba Science City, Japan" (Pickering 1992). I will explain in chapter eight why this work is important to my intellectual development, but there is no doubt that Traweek's, work on Japanese and USA physicists has left an indelible mark on my consciousness. Traweek, as an anthropologist, spent 30 years learning the culture of Japanese physicists, but her work is about so much more than what Japanese physicists do. Her work is about knowledge construction and power struggles. The male Japanese physicists had to deal with their own struggles for legitimacy in order to gain respect from Western physicist, but women, and not just Japanese scientists have to overcome much more than ethnic biases in the sciences. Traweek tells the story of trying to get scientists to recognize that they construct knowledge and as knowledge constructors they tell each other stories to convey this knowledge in certain, legitimized ways. Traweek's work is another example of how science is too important for non-scientists to ignore, and since that first excursion into her creative mind, when Traweek publishes something new, I read it.

Londa Schiebinger has done important work in feminist science as well. She suggests that there are four phases in the consciousness of those interested in feminism and science. There was initially the push to make sure great women scientists were included in any discussion of great scientists. This proved that when given the chance women could be innovators and trailblazers. Next there were attempts to uncover ways in which gender shaped what was being done in the sciences. In other words, gender shaped who decided what important research was and how that research should be done. This became a matter of exposing how such powerful terms as objectivity, subjectivity, value neutrality, and universal laws were defined and created. Then there was a movement to "uncovering sexism in research results. This crucial step unmasked the claim that science is gender neutral and underscored how gender inequalities have been built into the production and structure of knowledge" (Schiebinger 2004, 234). Finally, there is a shift from feminism and science to feminist science in which the main question is self-reflexive of those who do feminist science and of those, in general, who do science. "Scholars have begun," Schiebinger (2004, 234) suggests, "to ask how feminism—conceived both as a political and social movement, and as an academic perspective— has brought new questions and priorities to the sciences." This last point, in my opinion, is a major milestone in science studies because the debates are no longer about how does gender bias shape science and how do male scientists construct stories to deny they construct stories. The issue is now how can feminist science shape research agendas that shape the knowledge everyone uses to make policies, live healthy lives, and thrive in a democratic world. In another article Schiebinger (2003, 861), with the help of an archeologist, Margaret Conkey, and a physicist, Amy Bug, summarizes what some of the principles feminist science might embody including "democratizing research, eliminating research that leads to exploitation of nature or other humans, resisting explanations stripped of social and political context…acknowledging our values and beliefs, being honest about out assumptions, being responsible in our language." This is a research agenda I can endorse. Nowhere in these principles does it say that science is not an empirical endeavor or that feminism is in conflict with protocols of establishing facts. Feminist science is concerned with how one gets to the moment of establishing facts and declaring something a "Truth." For Feminist scientists how a scientist gets to a "fact" or a "truth" is just as important as "discovering", "declaring," or "uncovering" a "fact." "Doing science", Schiebinger (2003, 862) thinks, "as a feminist means mainstreaming politically engaged gender analysis into all aspects of science–its institutions, theories, practices, priorities, and policies. Doing science as a feminist means, as [Patty] Gowaty phrases it, 'to modify the rhetoric of the women's movement into stable hypotheses' for use in science." The question, therefore, is not only how can science become more inviting to female scientists, but also how can feminist concepts, philosophies, and histories define what scientists do? Such a question in and of itself marks a shift in power from the assumption that science is the ultimate form of knowledge and the arbiter of what is truthful and correct to acceptance that science is limited in its truth claims and can do much better in serving the world by opening its work up to feminist thought.

In my own intellectual experiences, the most influential scholar of feminist science is Donna Haraway. Most know her because of her famous 1985 cyborg manifesto in which she declares that she would rather be a cyborg than a goddess. Since this work is so well known I want to focus on some of her other important work in science studies. One of the major themes that Haraway, along with other feminist scholars, has challenged is the notion of objectivity or what Steven Shapin and Simon Schaffer refer to as the modest witness. The modest witness in early modern science became a way of establishing a fact that can be used to conduct more empirical work in order to construct and authorize a scientific agenda. Robert Boyle in the 1660s is considered to be the father of the modest witness and Shapin and Schaffer tell his tale of success to establish his approach to science. Like anything in real life, the legend of the objective, trustworthy, scientist and the struggle to establish a protocol for establishing facts are two different affairs. Boyle firmly believed that there was only one way to establish a fact and that is through empirical experimentation. As Shapin and Schaffer (1985, 65), note "The ability of the reporter to multiply witnesses [more witnesses, more likely a fact will eventually be established] depended upon readers' acceptance of him as a provider of reliable testimony." Could Boyle be trusted by others to report truth? Boyle believed there were a few ways in which trust could be established. One way was to demonstrate one's uncompromising honesty by "reporting all experimental errors" (Shapin and

Schaffer 1985, 65). Another way was to make sure the "correct" eyes saw and corroborated what the scientist said happened. A third way was to insure that the scientist was a modest man who did not bring attention to himself but to the reality being revealed; the fact being unveiled. In order to insure all of this those who were permitted access to the laboratory had to be limited. Because Thomas Hobbes doubted the veracity of Boyle's approach to establishing a fact, Hobbes was not allowed into Boyle's demonstrations, as a result Hobbes was not deemed a reliable witness and not admitted into the Royal Academy of Science. Hobbes did not believe that Boyle's experimental work could be construed as a philosophy. More importantly, "he refused to credit experimentalists' claims that one could establish a procedural boundary between observing the positive regularities produced by experiment (facts) and identifying the physical cause that accounts for them (theories)" (Shapin and Schaffer 1985, 111). The artisans who helped Boyle construct his air pump were not banned from the demonstrations because their expertise was necessary for a successful experiment. They, however, had the wrong eyes. Theirs were unreliable because they were nor aristocratic. Boyle's eyes could be trusted because he did not do scientific work to be paid, he did it as a sense of duty for the glory of the king and in the name of truth. Boyle was modest, but the artisans were immodest in motive and disposition. Women were also banned from baring witness. They asked too many unimportant questions like why does an animal have to be sacrificed in order to affirm that the air pump was completely sealed? It seems seventeenth century women were already living Schiebinger's feminist science. As a result of these questions, Boyle decided he had to ban women too so he only told his male students when he would perform his experiments so they could attend. Does this mark the beginning of the long decline of women's interest in science? With Boyle's attempt to establish the modest witness as a pillar of experimental science, politics (the exclusion of Hobbes from the Royal Academy of Science), gender relations (banning women from experiments), and class relations (discounting the testimony of artisans) already played a role in creating modern science.

For Haraway the modest witness is part of science's history of creating "a culture of no culture" (a term she borrows from Traweek). In this culture of science that denies it has any culture at all, the modest witness, "the legitimate and authorized ventriloquist for the object world, adding nothing from his mere opinions, from his biasing embodiment. And so he is endowed with the remarkable power to establish the facts" (Haraway 1997, 24). When the modest witness speaks, he is not speaking, rather it is "nature" that is speaking in all its power and truthfulness. For the modest witness to speak would corrupt him and nature. This is why Hobbes, artisans, and women were banned from "witnessing" the experiments even if they were present. They were not qualified or expert ventriloquists. When they spoke, they let their emotions and humanness get in the way. So they were told and so too many believed this is how scientists worked. Haraway (1997, 24) seeks to take the modest witness and "queer" him in order to make sure that "a more corporeal, inflected, and optically dense, if less elegant, kind of modest witness to matters of facts to emerge in the worlds of technoscience." Haraway, like other feminists, is not interested in emasculating the male scientist. She is trying to make the modest witness more human. Once the modest witness representing the culture of no culture is revealed as a cultural being then the skeptics with credentials like Hobbes, the engineers like the seventeenth Century artisans, and women are valued more. Their eyes begin to see and more importantly to count. It is important for scientists to try and experience how "nature sees," but a queer modest witness is not an exclusive men's club. It is not the Augusta National Country Club after all; it is open to all people willing to study science and the ways of this strange but important culture functions.

Calling out modest witnesses, is not Haraway's only contribution. Her work is steeped not only in gender studies, but she remains very much interested in postcolonialial thought, race, and companion species. Although, race and non-human animals are very important dimension of science studies, I will focus my attention on postcolonialism and what feminists of science add to the discourse. In the discussion of postcolonialism and science, the topic of objectivity and its abuse against colonialized people is an important issue. Haraway, along with Sandra Harding, are major players in the debates.

"A Cyborg Manifesto" is not Haraway's only famous essay. Another important essay that shaped my thinking about science is her essay "Situated Knowledges: The Science Question in Feminism and the Privilege of Partial Perspective." As she does in her book on the modest witness, in this essay she takes on two competing interpretations of objectivity in order to construct her own vision. The traditional notion of objectivity has already been partially covered in the discussion of the modest witness. To be objective, in a traditional notion, is to let the "facts" speak, to write in the third person, and deny that any personal opinion or emotions have influenced the development of a "fact." To be objective is to be value free, apolitical, and unemotional, that is, without bias. In a traditional approach, scientists are not to concern themselves with political debates, ethical dilemmas, possible controversial outcomes and uses of scientific creations. The scientist is to merely concern himself with the doing of science and let others sort out how the science might be used. To get involved in political, ethical, or policy debates is to compromise the integrity of the scientist. This tradition apparently travels as far back as Archimedes since Serres (2012, 68) reports that as soon as the Roman's defeated the city state of Syracuse, a city he used his talents to build defenses, Archimedes offered his services to Romans: "Always first, continually victorious, a collaborator, he joined the camp of the victors, the only country he had ever known.". Archimedes, as a scientist, was not interested in getting involved in geo-political struggles. He was only interested in doing science.

Then there were the social constructivists. Science, when it is guided by a blind principle of objectivity and governed by the modest witness, is an ideology and as a result can only serve as a dangerous guide to power. Traditional science also is a mere smoke screen masking what really goes on in the laboratories and field work of scientists. To follow the myths that scientists tell themselves and others is to get lost and distracted from how scientists create, discover, classify, interpret, and pro-claim. "From this point of view," Haraway (1991, 184) contends the social construc-tivists or the sociologists of science believe, "science—the real game in town, the one we must play—is rhetoric, the persuasion of the relevant social actors that one's manufactured knowledge is a route to a desired form of very objective power."

Both of these approaches leave Haraway wanting. "So, I think my problem and 'our' problem is how to have simultaneously an account of radical historical contingency for all knowledge claims and knowing subjects…and a no-nonsense commitment to faithful accounts of a 'real' world, one that can be partially shared and friendly to earth-wide projects of finite freedom, adequate material abundance, modest meaning of suffering, and limited happiness" (Haraway 1997, 187). Haraway wants to recognize three things at once. She wants to embrace the power and importance of science to shape meaning and truth without the gamesmanship to monopolize these terms and then construct a culture in which culture is denied. She insists on recognizing that women, colonialized people, people of color, working class people, and many others have suffered from science because scientists and their ideological spokespersons refused to admit that science is contingent; never free of history, human beliefs, politics, or culture. Haraway delivers, like Schiebinger, a different world than modern science mytholigizes. She wants a science that is based in "reality" but one that is not in competition with the earth and threatening its survival. She also envisions a science that is not biased towards the West but instead takes into account different perspectives from different cultures throughout the world.

This is where the work of Sandra Harding is important. Harding simultaneously with Haraway builds an argument against traditional objectivity. Harding is a master diplomat. She is able to take the traditional Western, male concept of objectivity and use it against this very tradition. A common complaint levied against feminists and the sociologists of science Haraway mentions, is that they are relativists who are undermining not only the authority of science but their power to establish truth because now anything goes and since everyone has an opinion no one can say who is right or wrong. This is, of course a caricature and over-reaction to criticism about traditional science. Relativism is an untenable position because, like Boyle's modest witness or the traditional apolitical, value free scientist, Haraway notes, it is a god-trick. To take a relativist position is to proclaim that one knows every possible position and can proclaim them all equal. Relativism is to replace a universal claim with a universal claim. Feminists of science like Harding do not take a relativist position. What Harding has done with her position of strong objectivity is to redefine the meaning of objectivity. Traditional objectivity is a weak form of the concept because it does not capture all the players in the scientific process. Any scientist who is really interested in doing sound science would take into consideration all the important, human and non-human actors in the scientific process. A scientist who took into consideration all the groups impacted by their work would be practicing strong objectivity. For instance, if a scientist were doing field experiments in a rain forest or mapping potential sites for new oil drilling, the scientist would take into account the impact this work might have on all of the people and other animals who might benefit or suffer from the scientific work. Harding's strong objectivity would revolutionize how science is done because it would change how so much of science has been done in the past and, unfortunately, still is being done in the present. "The postcolonial science and technology histories," Harding (1998,

46) writes, "record case after case where scientific research clearly was not intended to increase 'human' freedom and general social welfare for the peoples Europeans encountered.". If Europeans, or anyone else doing science wished to conduct legitimate, sound, respectable, and acceptable research they would have to end this historical trajectory and begin to take into account the impact their research would have on indigenous peoples and other non-human animals.

2.8   What Is an Economics of Science?

When the term economics of science comes to mind often the thought is how economics shapes science. If this were the case, then the economics of science would be nothing new at all. Since the invention of the idea of science in ancient Greece, economics has always been an aspect of science. The reason Archimedes, as Serres tells the story, was so interested in the victors is he knew they could afford his talents and fund his work. Boyle considered himself a reliable witness of facts because he had independent economic means to support his research, Galileo wished to impress the Medici family because more than the university they could fund his projects. The list of scientists seeking government funding, private philanthropic support, or even on-line democratic support from people interested in donating ten dollars to a research cause continues to grow. When the economics of science is discussed it refers to two different meanings that shape science. The first meaning refers to the use of economics to explain how science is done, dictate policy, discipline science, and convert knowledge into another commodity that can be thought of mathematically and exchanged like anything else on the "free" market. The Second meaning, and most important for this book, refers to the ways in which economic metaphors are used to explain science. Once economic metaphors began to be used more frequently in scientific discourse, it served as an external reminder of how economic thought has come to dominate the thinking of scientists. Metaphors matter and I would like to submit they embody the meanings of research trends and dominant paradigms. The economics of science, therefore, does not mark a break from science studies, but a continued effort to understand how science is done and how science impacts society. The economics of science is an important new step in the history of science studies because it is the first major critique of the impact of economic thought on science. If a sustained critique of economics is not developed, the monological and myopic dominance of neoliberal economic thought will prevail with dire consequences for thought in general and particular for science studies. The rise of neoliberal thought and Homo economicus also threatens the vitality and integrity of the university, the idea of a living democracy, and the necessity of looking at science as a sociological and cultural phenomenon. To write it more bluntly curriculum scholars, neoliberal thought threatens your vitality as well!

In his first major work looking at the rise of neoclassical economics during the middle and late Nineteenth century, Philip Mirowski (1989, 108) wrote that economists "have consistently lagged behind physicists in developing and elaborating metaphors; they have freeloaded off of physicists for their inspiration, and appropriated it in a shoddy and slipshod manner." He also demonstrates that the shoddiness of economists who tried to legitimize economics as a science in the end did not matter. Their use of physics to rationalize their economic theories was wrongheaded from a nineteenth century physicists perspective, but it worked. Economics gained a foothold through the use of the king of the knowledge hierarchy to gain legitimacy as a field of knowledge.

Since then the relationship between economics and the sciences have changed and in many ways reversed positions. In his seminal article "Blurred Boundaries: Recent Changes in the Relationship Between Economics and the Philosophy of Natural Science," D. Wade Hands (1994, 751) suggests like Mirowski that economists "writing in the field of 'economic methodology' would simply 'borrow' various arguments from the philosophy of natural science and then 'apply' those arguments to economics." By the end of the twentieth century this changed. "It is no longer the case," Hands (1994, 751) "that ideas flow exclusively one way. Economic methodologists are now considering a much wider range of philosophical sources, and philosophers of natural science have started to include economics in their philosophical investigations." In this article Hands (1994, 752) looks at four philosophers of science and how they use "economic ideas…in areas of philosophical inquiry where traditional epistemic concerns continue to be the primary focus." In other words, economic thought is serving as heuristic and foundational rationale to structure the thought found in a growing number of works in the philosophy of science. For example he relies on the work of Nancy Cartwright a leading philosopher in science studies. To no one's surprise Cartwright suggests something that is now typical in science studies: "Empirical facts are not simply 'discovered'; they are the product of negotiation by many agents and the result of the pragmatic rationality and the tacit local knowledge that characterizes experimental practice." (Hands (1994, 759) What Hands shows though is Cartwright built her argument on how science is done using John Stuart "Mill's view of laws in economics; it is a clear case of the philosophy of natural science borrowing directly from economic methodology." Economics as a model for thought, heuristic devices for constructing arguments, and the foundation for philosophies have become common place justifying a careful study of economic thought as a science and as a foundation for science. Economics has become a new dominant paradigm and as a result it is important to understand what the assumptions are regarding the impact of this thought and the use of certain metaphors to rationalize policies and notions of truth. The next two chapters deal with the rhetoric of economic thought and then in chapter five I will focus of the economics of thought, neoliberal thought, democracy, and becoming an expert in order to challenge the dominance of neoliberalism and its hijacking of neoclassical economics to force its ideological policy agenda onto the people of the world thereby openly challenging the roots of democracy.

References

Babich, B. (1994). *Nietzsche's philosophy of science: Reflecting science on the ground of art and life*. Albany: SUNY Press.

Biagioli, M. (1993). *Galileo courtier: The practice of science in the culture of absolutism*. Chicago: University of Chicago Press.

Bourdieu, P. (1975). The specificity of the scientific field and the social conditions of the progress of reason. *Social Science Information, 14*(6), 19–47.

Casti, J. (1998). *The Cambridge quintet: A work of scientific speculation*. Reading: Perseus Books.

Casti, J. (2003). *The one true platonic heaven: A scientific fiction on the limits of knowledge*. Washington, DC: Joseph Henry Press.

Dyson, F. (1997). *Imagined worlds*. Cambridge, MA: Harvard University Press.

Dyson, F. (1999). *The sun, the genome, and the internet: Tools of scientific revolutions*. New York: The New York Public Library, Oxford University Press.

Eisenstadt, S. N. (1968). *Max Weber: On charisma and institution building*. Chicago: University of Chicago Press.

Feyerabend, P. (1988). *Against method*. London: Verso Press.

Fox Keller, E. (1983). *A feeling for the organism: The life and work of Barbara McClintock*. New York: W.H. Freeman and Company.

Fox Keller, E. (1995). *Refiguring life: Metaphors of twentieth-century biology*. New York: Columbia University Press.

Gigerenzer, G., Swijtink, Z., Porter, T., Daston, L., Beatty, J., & Krüger, L. (1989). *The empire of chance*. Cambridge: Cambridge University Press.

Hands, D. W. (1994). Blurred boundaries: Recent changes in the relationship between economics and the philosophy of natural science. *Studies in the History and Philosophy of Science, 25*(5), 751–772.

Haraway, D. (1991). *Simians, cyborgs, and women: The reinvention of nature*. New York: Routledge Press.

Haraway, D. (1997). *Modest_Witness@Second_Millennium. FemaleMan©_Meets_OncoMouse™: Feminism and technoscience*. New York: Routledge Press.

Harding, S. (1998). *Is science multicultural?: PostColonialism, feminisms, and epistemologies*. Bloomington: Indiana University Press.

Hayles, N. K. (1984). *The cosmic web: Scientific field models & literary strategies in the 20th century*. Ithaca: Cornell University Press.

Hayles, N. K. (1990). *Chaos bound: Orderly disorder in contemporary literature and science*. Ithaca: Cornell University Press.

Hayles, N. K. (1999). *How we became posthuman: Virtual bodies in cybernetics, literature, and informatics*. Chicago: University of Chicago Press.

Kant, I. (1798/1979). *The conflict of the faculties*. Lincoln: University of Nebraska Press.

Kuhn, T. (1962/1970). *The structure of scientific revolutions*. Chicago: University of Chicago.

Latour, B. (1987). *Science in action*. Cambridge, MA: Harvard University Press.

Latour, B. (2004). *Politics of nature: How to bring the sciences into democracy*. Cambridge, MA: Harvard University.

Latour, B., & Woolgar, S. (1986). *Laboratory life: The construction of scientific facts*. Princeton: Princeton University Press.

Lenoir, T. (1997). *Instituting science: The cultural production of scientific disciplines*. Stanford: Stanford University Press.

Lewontin, R. C. (2000). *The triple helix: Gene, organism, and environment*. Cambridge, MA: Harvard University Press.

Mirowski, P. (1989). *More heat than light: Economics as social physics, physics as nature's economics*. Cambridge: Cambridge University Press.

Mirowski, P. (2004). *The effortless economy of science*. Durham: Duke University Press.

Newman, J. H. (1852/1996). *The idea of the university*. New Haven: Yale University Press.

Nietzsche, F. (1974). *The gay science* (trans: Walter, K.) New York: Vintage Press.

Nowotny, H., Scott, P., & Gibbons, M. (2003). Introduction 'mode 2'revisted: The new production of knowledge. *Minerva, 41*, 179–194.

Pestre, D. (2003). Regimes of knowledge production in society: Towards a more political and social reading. *Minerva, 41*, 245–261.

Pestre, D. (2004). Thirty years of science studies: Knowledge, society and the political. *History and Technology, 20*(4), 351–369.

Pestre, D. (2007). The historical heritage of the 19th and 20th centuries: Techno-science, markets, and regulations in a long-term perspective. *History and Technology, 23*(4), 407–420.

Pickering, A. (Ed.). (1992). *Science as practice and culture*. Chicago: University of Chicago Press.

Pickering, A. (2005). Decentering sociology: Synthetic dyes and social theory. *Perspectives in Science, 13*(3), 352–405.

Rajan, K. (2006). *Biocapital: The constitution of postgenomic life*. Durham: Duke University Press.

Readings, B. (1996). *The university in ruins*. Cambridge, MA: Harvard University.

Rheinberger, H.-J. (1992a). Experiment, difference, and writing I: Teaching protein synthesis. *Studies in History and Philosophy of Science, 23*(2), 305–331.

Rheinberger, H.-J. (1992b). Experiment, difference, writing II. *Studies in History and Philosophy of Science, 23*(3), 389–422.

Rheinberger, H.-J. (2003). 'Discourse of circumstance': A note on the author in science. In M. Biagioli & P. Galison (Eds.), *Scientific authorship: Credit and intellectual property in science* (pp. 309–323). New York: Routledge.

Schiebinger, L. (2003). Introduction: Feminism inside the Sciences. *Signs: Journal of Women in Culture and Society, 28*(3), 859–866.

Schiebinger, L. (2004). Feminist history in colonial science. *Hypatia, 19*(1), 233–254.

Serres, M. (2012). *Biogea*. Minneapolis: Univocal Press.

Shapin, S., & Schaffer, S. (1985). *Leviathan and the air-pump: Hobbes, boyle, and the experimental life*. Princeton: Princeton University Press.

Snow, C. P. (1998). *The two cultures*. Cambridge: Cambridge University Press.

Part I
Interlude One: The Fault of Noreen Garman

My title is a play on the subtitle of the French Philosopher Bernard Stiegler's first translated book into English, *Technics and time 1: The Fault of Epimetheus*, in which he tells the tale of Prometheus and Epimetheus and how humans were always connected to and with technology. Epimetheus always came after. His brother took the lead, after all he was Prometheus, but Epimetheus wanted to show to his brother that he could handle the responsibility given to the one who came first by the gods to hand out the talents and powers to all the living creatures. When it came time to bestow some abilities onto the humans Epimetheus ran out. He had failed his brother and his brother the gods. There would be Hades to pay for such lack of foresight. Never give a job to someone who always comes after, comes too late, thinks too slowly. Powerless and facing the uncertainties of the world because Epimetheus's realization did not come quickly enough, humans from the start had to fend for themselves by inventing their own abilities. They had to utilize their own language and hands to create what they needed. This was, for Stiegler, the first interactions between humans and technology, and it happened right at the beginning. Humans are technological from the start.

Why am I so critical of methodology, especially the stifling non-thinking found in most activities of education? Why do I resist all attempts to codify research protocols? Why do I instinctively reject any notion that sets a method into steps and pretends it guides any search for truth or anything empirical or factual? In a few words: it is the fault of Noreen Garman. It was Noreen Garman who introduced me to the history of rhetoric and the rhetorical turn of the 1980s and 1990s. My first class I took with her was called of all things qualitative research. One of the many items she had us read was an article written by Margaret Marshall and Loren Barrit titled *Choices Made, Worlds Created: The Rhetoric of AERJ*. What a fascinating article, published in of all places *AERJ*, the American Educational Researcher Journal, the official journal of the American Educational Research Association. What a dreadfully dry journal. Still to this day it is stuck in the positivist realm in which statistics, traditional anthropology, and modest "mixed methods" serve as the dominant paradigms and deep seated ideology. This article looked at the rhetoric of the journal and confirmed those who had an inkling of suspicion of the deep biases of the journal towards only a certain type of research. Marshall and Barritt (1990, 593) noted the articles in AERJ "conform to a predictable form." Anything that went outside of the boundaries, tightly guarded by the methodology police who know the protocol and know interlopers and intruders when they see them, was not legitimate research, never scholarship, and an abomination to the science gods. If one were to pick up recent issues of AERJ and then explored a few more representative issues, the newcomer reader would be worm-holed back to Post War World America where Positivism and empirical Social Science reigned. One would never get the sense that intellectual debates raged in the 60s over Poststructuralism and Feminism, Postmodernism in the 90s or Posthumanism or New Materialism in the early Twentieth-first Century. These intellectual traditions and scholarly approaches simply do not exist in the minds of the AERJ faithful. When the idea of innovation and compromise is considered to be mixed methods, one without doubt enters into an intellectual wasteland where researchers are intellectually intimidated to venture beyond the norm. The binary of quantitative or qualitative methods with an Aristotelian golden mean of mix methods in the middle still holds firm in the (non) thinking of too many in the fields of education.

After I read the *The Rhetoric of AERJ* I asked Noreen to feed me some more. She had piqued my curiosity. I was convinced of the intellectual limitations of traditional scholarship before I entered into her class, but no doubt she convinced me that my suspicions were well founded. I needed more to sustain my new curiosity. Because of who she is, she fed me. Noreen introduced me to the scholars who were reintroducing rhetoric to USA intellectual debates. She did not just guide me to rhetoricians. When she found out in 1994 that I was offered a position at LSU-Shreveport, Noreen hugged me and said two things: "I am so happy you found a job, we were worried our students would never find one and make sure you look up the two Bills." That was Bill Pinar and Bill Doll. I indeed did. Bill Pinar supported me in getting graduate status in his department in Baton Rouge so I could teach doctoral courses, just as Joe Kincheloe did in the 1980s, in Shreveport. Bill Doll came to my class one semester and spoke to my students and it was afterwards at dinner that he suggested I attend Bergamo which at the time was being held in Tennessee. I did and thus began my career in Curriculum Studies. This too is the fault of Noreen Garman.

In the two chapters that follow I name names. I share with the reader those who were leading the intellectual resurgence in the late Twentieth and early Twentieth-first Centuries, trying to return rhetoric to the realm of scholarly inquiry. I move from the rise of rhetoric in Ancient Greece and Rome through the early modern period and the first attempt to banish rhetoric from human inquiry, rhetoric's return to the human and social sciences, and I end with a discussion of the rhetoric of science. In the second chapter of this section I do some applied rhetorical analysis with such terms as data, models, and statistics. In both chapters I use economics as the back drop because economics may be glibly viewed as the "dismal science," it is a science and, along with biology, the field that reigns supreme in our unfortunate continuation to construct hierarchies of knowledge. Deniers of climate change certainly question the outcomes of the sophisticated models of climate scientists but strangely they never question the equally complicated models, and the assumptions embedded in them, of economists and mathematicians. It is just one example of the rhetoric of hierarchical knowledge and where scholars and non-scholars place their faith. Of course all of this is the fault of Noreen Garman, but nothing in these chapters that are deemed out of bounds or incorrect are not. She helped create me but, like any good professor, she did not dictate the boundaries from which my mind could roam. I did.

References

Stieger, B. *Technics and time, 1: The fault of Epimetheus*. Stanford: Stanford University Press.
Marshall, M., & Barritt, L. (1990). Choices made, worlds created: The rhetoric of AERJ. *American Educational Research Journal, 27*(4), 589–609.

Chapter 3
Homo Economicus, Rhetoric, and Curriculum Studies

Curriculum Theorists have rightfully criticized Marxist thought as being too reductionist. Marx's classic example of reductionist thought comes from his early work with Friedrich Engels in which they proclaimed:

> In the social production of their life, people enter into a specific, necessary relations of production…correspond[ing] to a specific state of development of their material forces of production. The totality of these relations of production forms the economic structure of society the factually existing base, on which a juridical and political superstructure is raised…The mode of production of material life determines the social, political, and intellectual process of life in general. (Sperber 2013, 400)

For Marx and Engels economics determined one's life trajectory, it was the major foundation for all that was to come and to shape who one would and could be. Familial traditions, political barriers, societal discrimination, cultural influences, and personal habits were all subordinated to and determined by an economic foundation. Critics of Marxism were right to point out the contingencies of life and how sometimes, in spite of one's background and economic realities, individuals can transcend their class position and economic condition. Sometimes economic foundations did not matter, many times a person's culture, race, gender, sexual orientation, political affiliation, (non)religious beliefs, and other societal factors trumped economic realities. Sometimes economic foundations were subordinated to other conditions.

Nancy Folbre and Heidi Hartmann provide evidence to demonstrate how neoclassical economists, Marx, and the Marxist tradition ignore the many ways in which gender is a prevailing marker for shaping the reality of individuals. In their work they demonstrate how both neoclassical economists and Marx shaped women as non-economic entities with no ambitions in the public sector of economic exchange. Folbre and Hartmann (1988, 189) suggest "many neoclassical economists argue that women place a higher priority on the welfare of their family than on the level of their wages." Women in the model building imaginations of neoclassical economists are altruistically domesticated, reduced to doting soccer moms,

© Springer International Publishing AG, part of Springer Nature 2018   49
J. A. Weaver, *Science, Democracy, and Curriculum Studies*, Critical Studies
of Education 8, https://doi.org/10.1007/978-3-319-93840-0_3

worrywarts wondering if their son or daughter will get into the right preschool and never bothered about earning as much as their male counterparts at work.

Marxist economists do not fare any better. Folbre and Hartmann (1988, p. 191) point out that Marxists have subsumed many women's issues into class issues. "Thus, members of families," they suggest, "are assumed to have the same class membership and class interests as their male wage earner." What's good for Archie is exactly what Edith needs and wants. In this mind set working class women have no other needs then to tend to the family unit and if possible bring in a supplemental income. Under Marxism, women are erased and reduced to the family unit that is defined ultimately by economic realities.

The same could be said in regards to any other social ascriptive group as well, not just women. The concerns of Asian gay, non-believing men, Latina, straight, Roman Catholic women or a transgendered, European-American, Southern fundamentalists are reduced to their economic status. By questioning the economic superstructure critics in numerous fields including curriculum theory have broadened scholars' and non-academic's understanding of how people experience the world. Challenging the hegemony of Marxian economic reductionism has led to a theoretical bonanza in countless areas including the rise of identity formation, cultural studies, gender studies, gay and lesbian studies and other fields of knowledge.

Marx's reductionist tendencies were challenged in other areas as well. Although it may be hard to accept such a view today after the fall of the Soviet Union and their "sphere of influence", Marx was an optimist. He offered scathing criticism of nineteenth century capitalism and pointed out its many flaws, but he ultimately believed history was on the working class people's, the majority's, side. Marx had reduced the flow of history to an inevitable course that would mark the rise of the bourgeoisie, their eventual demise, and the ultimate rise of working people. This view of history embodies a great faith in the human spirit and individuals' desire to be free and prosperous. Unfortunately and fortunately, Marx was wrong. History does not and cannot guarantee progress, no matter who tries to define what progress might mean. There is no guarantee even the most corrupt, connected, and powerful of people such as the Koch brothers will succeed in implementing their vision of what they think the present and future should be. There is no guarantee that the futures of our children will be better than the present adult generation no matter how much parents may believe it to be so. I write that Marx's vision of history was unfortunately wrong because it does not take much searching to see how much suffering an uncertain future brings. A prosperous family 1 year can be a foreclosed, homeless family the next year as their bubble burst because of any number of reasons including but not limited to predatory lenders, a corrupt political system, crony capitalism in which risk is socialized and profits privatized, shady accounting, or inflated income projections. At the same time, it is fortunate that Marx's views of historical movements were wrong simply because a life filled with guarantees is no life at all. It is a script, played out not by actors but by automatons. I believe there should be some guarantees such as access to an education, housing, and health care, but a life without the risk of failure or the potential to not succeed is lifeless.

Reductionism has emerged from Marx and Marxism in another important way too. Marx himself has been reduced to a caricature of himself. Marx who never lived to see the twentieth century is most associated with that century than he is with his own. Marx is viewed as a failed economic, political, and philosophical thinker because he has been connected to the failures of the Soviet Union, China, Vietnam, Cuba, and Eastern Europe. The failures of these dictatorships is no more the fault of Marx as are the failures of crony capitalism the fault of Adam Smith. Yet, Marx has been reduced to political and economic Marxism born in the twentieth century. In an interview on his new biography of Karl Marx, Jonathan Sperber said one of his goals for writing the book was to return Marx to the nineteenth century. This is exactly what needs to be done if we have an interest in understanding the complexities of Karl Marx. This also means those of us who are interested in creating an alternative vision to neoclassical economics will have to chart our own course, leaving the nineteenth century behind and addressing the uniqueness and specifics of the twenty-first century using our own circumstances, imaginations, and possibilities. This will not be easy I am sure.

There is one more way Marx is a victim of reductionist thought. The phrase "opium of the people" is often bantered about in religious and political circles but it is so often misunderstood. Viewed by the uninformed and unmotivated as proof that Marx was an atheist who thought religious people were foolish, it is actually a phrase filled with empathy and compassion. Marx was an atheist no doubt, but Marx's famous phrase has to be taken into context to understand what he was saying. The full passage reads:

> Religious suffering is at the same time an expression of real suffering and a protest against real suffering. Religion is the sigh of the oppressed creature, the sentiment of a heartless world, and the soul of soulless conditions. It is the opium of the people.

> The abolition of religion as the illusory happiness of men, is a demand for their real happiness. The call to abandon their illusions about their condition is a call to abandon a condition which requires illusions. The criticism of religion is, therefore, the embryonic criticism of this vale of tears on which religion is the halo. (Smelser 1973, p. 14)

Marx sees suffering people all around him and understands why people turn to religion. Religion helps them deal with real pain. Religion is a pain reliever, a way to run from a heartless reality and soulless people who scorn and persecute the poor. At the same time, religion is merely an illusion because many times the religious do nothing to alleviate the pain, they only create illusions. The religious do not tend to the causes of tears; they only cover them up and then pretend they did something saintly and good. Marx's criticism of religion is accurate and something necessary for anyone who calls themselves religious. Instead of being a vale, a source of illusory happiness, or a front for the heartless, religion needs to become something real to suffering people. If not, then religion is exactly what Marx said it was. Those who reduce Marx's comments to proof positive that Marx was an atheist are speaking the obvious and avoiding the reality of religious influence on people. This is one major reason why so few people in the world actually trust religious institutions anymore. The self-anointed haloed ones should have read their Marx more carefully.

I begin this chapter with a lengthy look at the reductionism and reduction of Marx because a result of the criticism of Marx and Marxism has been the rise of new, more dynamic critical thought and new approaches to theory, but it has also resulted in the abandonment of critical commentary on economic thought. As a result of this abandonment what we have seen is the continuation of reductionist thought. While Marxist reductionist thought has been discredited, there remains no real active criticism of the one true reductionist meta-narrative that remains in the postmodern world: Homo Economicus. In an intellectual climate in which contingencies of thought and uncertainty of possible futures reign supreme and promise complicated futures, everything has been reduced to neoliberal economics.

It is not hard to find everyday proof of how life has been reduced to Homo Economicus. Clichés are words or phrases that express nothing. They are a form of communication that communicates nothing. We hear clichés every day and they are usually sports related:" Its fourth down and time to punt," "it's the bottom of the ninth," or "there is no I in team." In regards to Homo Economicus clichés are very revealing. How often you do you hear someone say "the bottom line is…" This clichés, of course, economic in origin, is a way of saying nothing with words. It is a trite and concise way of saying "I have better things to do with my life than to speak with you" or "I am in a hurry so I want everything in a quick, concise format." The language of Homo Economicus also expresses our values. When I see silly suction placards on minivans that read precious cargo onboard I realize it is meant to be some kind of expression of love but it is a trite way of reducing children to objects. When my children were born I did not proclaim that my special package arrived! Yet so many people refer to their children as objects. I often hear parents proudly proclaim that their children are the most precious thing they have. Children are not commodities, yet Homo Economicus reduces children to things. When our precious cargo grow up, go to college, and look for employment often adults tell the young to go and "sell themselves." This cliché sounds more like a plea for our children to find a way to repay their student loans via prostitution rather than sage advice on how to find employment. Another way Homo Economicus shows its ugly clichéd head in life is when adults tell young people that they need to get a good education in order to get a good paying job. Often I hear something like this: "Did you know that a college graduate will earn 1.7 million more than the average high school graduate?" This reduces the learning experience to job training and this is exactly what universities tout when they proudly proclaim that most of their graduates are successfully placed in jobs of their chosen careers. When we dispense with this advice or proclaim these results, we are turning learning and knowledge creation into an economic transaction. Even more fundamentally, we are eliminating joy from life, exiling the imagination and creativity from human experiences. As a result of the reductionist meta-narrative of Homo Economicus humanity has been reduced to an economic equation (i.e., rational man model, general theory of equilibrium) and a capital quantity (i.e., when we give x number of state dollars to higher education what is the return on "our investment?").

In this section of the book, I submit that curriculum studies scholars specifically and intellectuals in general have to reclaim a critical voice in discussing economic matters. The meta-narrative called Homo Economicus needs to be challenged. There continues to be cutting edge research on social class issues such as Jean Anyon's work or Michelle Fine and Lois Weis's work. This work is noteworthy and important for numerous reasons. The type of scholarship I am referring to, however, goes beyond social class issues, identity politics, and other important topics curriculum studies scholars focus on today. I want us to begin anew a discourse and criticism on neoclassical economic thought. Neoclassical thought is more ideology than it is economic knowledge. The crushing reality of people's experiences in all facets of life prove every day that neoclassical thought is not prepared to deal with or explain health care, education, poverty, religious intolerance, class warfare against working people, the disparity of income, political corruption, and global ecological crises. The free market neoclassical economists proclaim as universal and infallible needs competition and economists are not equipped or trained to do it. While lobbyists like ALEC (American Legislative Executive Committee) buy politicians and write legislation at the state and federal level and power brokers shape public policy, the neoclassical minions talk about freedom of markets and choice. There is no freedom or choice when incomes are in constant decline for working people while tax burdens increase and tax systems become regressive not progressive. There is no choice when someone has to choose between health care and food. There is no freedom when individuals can vote but the only two choices are handpicked by the Koch brothers or some other well connected billionaire who views democracy as a playground for their vanity. Neoclassical economic thought has strained under the pressures of reality, but there are no alternatives. It is our responsibility as curriculum scholars and other intellectuals to create them.

As Roger Brickhouse (2010, 6–7) notes some economics students recently at the University of Cambridge "circulated a petition criticizing the narrowness of economics and calling for debate over the foundations." French students followed suit and as did a group of Harvard students "wanting a curriculum that would be more critical of conventional ways of thinking." Convention so far is winning. It is time for curriculum scholars to reclaim a role of critic of economic theories and policies. The path I want to outline below is the use of rhetoric to reclaim our voice in economic debates. In this chapter I will outline why rhetoric is an appropriate avenue for curriculum theorists to take to begin the process of offering critiques of economic thought and policies and eventually presenting economic alternatives to neoclassical reductionism.

3.1 The Art of Rhetoric

I want to discuss in the sections to follow at least five reasons why it is important for curriculum scholars to reclaim an interest in the study of rhetoric as we return to a focus on economic thought and policy. This discussion will be very lengthy so I will not just list the five reasons why rhetoric is important but by the end of this chapter hopefully you will see the reasons why the path back to economic thinking is through rhetoric. I want to begin this discussion of rhetoric with Aristotle's treatise on *Rhetoric* because it is here where we find at least two very important reasons why the study of rhetoric is essential in order to enter into a conversation with economists. First, rhetoric is an art and everyone uses rhetoric. To apply rhetoric to economic thought reminds economists that their craft like all other crafts is an art as much as it is a science. When we discuss the importance of modeling in economic thought we will see how artistic economists are. Most economists, like most people in general, deny the importance of rhetoric in their craft. What this denial shows is that the good life is no longer a goal for most people and life, any kind of life, is good enough. To see what is lost when not focusing on the question of the good life and ignoring the importance of rhetoric in creating the good life is why it is important to begin with Aristotle, work through other rhetors such as Quintilian, Saint Augustine, Bacon (yes even someone like Francis Bacon saw the importance of studying rhetoric) Coleridge, I.A. Richards, Kenneth Burke, Wayne Booth, the rhetoricians of science such as Alan Gross, Dalip Parameshwar Gaonkar, and end with some important rhetoricians of economics in Deidra McCloskey, Arjo Klamer, and Philip Mirowski. All of these thinkers stress the art of all disciplines, and the artful use of rhetoric.

Even though it is important to start this section with a look at Aristotle's treatise on rhetoric it is a curious place to begin. One of Aristotle's major commentators today, Eugene Garver (1994) calls Aristotle's *Rhetoric* "a strikingly unreflective work" (8) whose "impact...has been negligible" (4). Not a resounding endorsement from one of the stronger readers of Aristotle's work. Aristotle's work is really best described as a manual and, really, in an age in which manuals and "how-to books" dominate the field of education, do we need another book like this? Although *Rhetoric* can be described as a manual it is a different type. Aristotle is trying to persuade his readers and pupils that in order to live in a vibrant democracy and live the good life rhetoric is an important art to master. While Aristotle's teacher viewed rhetoric the same way he viewed poetry and painting, a poor imitation of an imitation of truth therefore untrustworthy, Aristotle saw rhetoric differently. Rhetoric was a practical art of persuasion that had nothing to do with deception or distraction. Aristotle recognized that rhetoric could be used by selfish and dishonest speakers but the greatest defense against deception was to understand the art of rhetoric, to understand the common good and recognize when a speaker did not have the common good in mind. As a result, rhetoric for Aristotle was important in three areas: the political arena, legal settings, and eulogies or the praising of great people. If a student of rhetoric could master the skills of persuading in these areas then the common good would triumph over self-interest and misguided motives. Aristotle is important to understand when thinking about rhetoric because he recognized that rhetoric held an important and practical position in society. To create a sound government and society, one had to be a master of rhetoric, and more importantly, the audience, the ears upon which the speeches of politicians were targeted, needed to be well versed in understanding the practical impact of rhetoric. At the heart of the first two reasons to study rhetoric, the art of rhetoric and the good life, then is a very practical goal of creating a vibrant and lasting democratic society.

Aristotle also reveals a third very important reason to study rhetoric: rhetorical skills are not to be monopolized by professionals nor are these skills something that should be used in economic exchanges. To use rhetoric for economic exchanges is to demean the art of rhetoric and the good life, and it is to practice the art of deceit. "It is hard to see how something can be an art," Garver (1994, 7) warns, "without becoming a specialty for experts and therefore being for sale, exchange, or rent." This is a common theme that will dominate the study of rhetoric in a traditional liberal arts education for the next two millennia. Rhetoric is not a skill to be harnessed, bottled, and marketed. Rhetoric, instead, is a mark of a broadly educated, broad minded person. Rhetoric is "civic" that "aims at the internal end, and therefore governed by internal standards and values, while professional rhetoric, as a purely technical and rational dynamis, aims at an external end and so is ruled by ulterior motives....Professional arts aim only at external ends: the vulgar pleasure of the audience rather than the education of the practitioner" (Garver 1994, 50). To professionalize rhetoric, to codify it as a protocol rather than an art, is to reduce the pleasures of life to exchanges, economic or otherwise, and to turn the good life into just life. It is to let the "maxim that 'the customer is always right'" rule the day, reign over life like the grim reaper (Garver 1994, 31). My life then becomes merely an transaction with others in which I am daily tested to see what I can get from them and what they can get from me. The one with the most things is the winner. Sounds like a good definition of the neoclassical rational man model but a pathetic notion of what it means to live.

Garver notes the professionalization of rhetoric also leads to false security. When rhetoric is reduced to a protocol art disappears and in its place are its opposites, security and wealth. In the professional use of rhetoric the leaders are not evaluated on their abilities to sustain the common good but to secure wealth. Garver observes this reduced political activity to "the mere means to life, that is to wealth and security" (1994, 29). The more wealth one accumulates the more one needs assurances that their accumulations are well protected. The goal is no longer what the common good is but what will the leaders do to insure my goods are secure? Whatever plan protects my goods then that is the plan I will demand. My selfish needs are the only goal that should matter. We have stumbled upon another good definition of classical and neoclassical economics but hardly a definition of a good life.

Quintilian is considered the greatest teacher of rhetoric and lived in the first century of the Common Era and the height of the Roman Empire. Like Aristotle, Quintilian's aim was "the education of the perfect orator. This required two important goals. The first essential goal is that" one "should be a good person and consequently we demand of" him/her "not merely the possession of exceptional gifts of speech, but of all the excellence of character as well" (Wheelock 1974, 24). If the orator is not in possession of an excellent character then the likelihood that the orator will not serve the common good. It is the common good that ultimately is Quintilian's goal. "But above all" one "must possess the quality which Cato places first and which is in the very nature of things the greatest and most important, that is, he must be a good" person (Wheelock 1974, 148). There are two ways in which the teacher of rhetoric can insure the creation of a good person. First avoid the professionalization of rhetoric. "As soon as speaking became a means of livelihood," Quintilian observes, "and the practice of making an evil use of the blessings of eloquence came into vogue, those who had a reputation for eloquence ceased to study moral philosophy and ethics…became the prey of weaker intellects" (Wheelock 1974, 25). When rhetoric becomes a field of knowledge to serve economic concerns ("a means of livelihood") then the orator is subject to manipulative and selfish ends. Second goal is only serve the public, "if the powers of eloquence serve only to lend arms to crime, there can be nothing more pernicious than eloquence to public and private welfare" (Wheelock 1974, 148). Many in Quintilian's era and after believed that if the deceptive and selfish could develop rhetorical skill then they should not be taught. Quintilian objected to such logic and suggested that ignorance of an art is not the best defense against abuse but using rhetorical skills only for the public good and not confusing or blending the common good with private wishes was the best assurance against abuse. What is important to note here is how drastically different the notion of common good in Ancient Rome is from the twenty-first century. The common good, as will be the case for other thinkers in this chapter up to and including Burke and Booth, was not seen as an economic concept or goal. Economic needs were viewed as a source of manipulation and private desires; a disruption of the public good. Today economic needs are viewed as patriotic, democratic, and the essence of the common good. The orators today have persuaded the public that the economic wants of certain individuals is the common good.

Augustine was also a teacher of rhetoric. His influential autobiography, *Confessions*, can be read as an example of rhetoric in practice. In this book the addressee is God but in reality the targeted audience is the leaders of the fourth century Catholic Church. Augustine wished to prove to church leaders that he was not a heretical follower of the Manichees (who believed among other things that Jesus was not human) and worthy of a leadership position within the church. He was successful in his attempt since he eventually became the Bishop of Hippo. Augustine admitted, like Quintilian, that rhetoric could be used to manipulate and deceive, referring to them as "word spinners with nothing to say" (Augustine 2008, 112) and condemned the Manichean orator, Faustus, for fooling many to convert to his heresy "as a result of his smooth talk" (Augustine 2008, 73). Augustine even condemned his own use of rhetoric. "Overcome by greed myself," Augustine (2008, 53) explains, "I used to sell the eloquence that would overcome an opponent….Without any resort to a trick [in other words, he understood not to cross those boundaries of teaching evil] I taught them the tricks of rhetoric, not that should they use them against the life of an innocent man, [again he knew the moral boundaries of teaching oratory skills]but that sometimes they might save the life of a guilty person."

After these confessions, Augustine uses his skills to argue why the teaching of rhetoric is important, and as a result Augustine is able to remove rhetoric from the realm of pagan education and secure as a foundation for Western (Christian) Liberal Arts. Augustine argues that rhetoric informs the student what is true and good. To abandon the teaching of oratory skills would leave rhetoric to the word spinners and smooth talkers who would teach students how to become "friends of the Emperor" (2008, 143) rather than to seek the good life which is "the mere search for wisdom which should be preferred even to the discovery of treasures and to ruling over nations and to the physical delight available to me at a nod" (2008, p. 145). It is the teaching of rhetoric that allows us to ask the important questions of the good life and for Augustine the most important question to ask is: "What then do I love when I love my God? Who is he who is higher than the highest element in my soul" (2008, 185). By using rhetoric to make his case to the church leaders he not only secured a position for himself but also demonstrated how rhetoric was essential in any liberal arts education in Christian communities. Augustine believes rhetoric should be taught so everyone can understand how "the created order speaks to all" but is heard only by those who know how to "hear its outward voice and compare it with the truth within themselves" (2008, 184). Because of Augustine and Quintilian, rhetoric became an essential pedagogical tool to nurture the mind, cultivate the good life, confirm the common good, and prevent the manipulation of the common good for economic ends.

While Augustine solidified the role of rhetoric in the Western tradition of education, the modern period until the 1990s marks a period in which rhetoric is subordinated to an inferior position to "reason" and rhetoric and science are eventually separated. The most cited critic of rhetoric is of course John Locke. His diatribe against the teaching of any rhetoric is often cited as the symbol of modern disdain for rhetoric. Arjo Klamer and Thomas Leonard in their study of economic metaphors cites Locke's disdain for rhetoric. There is no discussion of the good life or common good in Locke's view of rhetoric. There is only criticism. For Locke to think plainly and "'to speak of Things as they are,…that the art of Rhetorick…all the artificial and figurative application of Words Eloquence has invented, are for nothing else but to insinuate wrong Ideas, move the passions and thereby mislead the Judgment'" (Klamer and Leonard 1996, 25). Rhetoric in the modern period is seen as an encumbrance to thinking, learning, and knowledge. Where rhetoric was the source of wisdom for Augustine, rhetoric is now viewed as the source of deceit and foolishness.

Even in the works of modern philosophers who advocate for the need of rhetoric in an educational approach concede that rhetoric is inferior to anything logical or rational. France Bacon in his treatise on the *Advancement of Learning* refers to the teaching of rhetoric as "the doctrine of ornament in speech, called by the name of rhetoric or oratory" (Bacon 1899, 176). This ornament also called eloquence, which of course is a great metaphor to describe rhetoric and its lowly position in modern thought, "is certainly inferior to wisdom." To prove his point about the separation between rhetoric and wisdom, Bacon resorts to a rhetorical strategy to reverse Augustine's argument. Where Augustine said rhetoric was the route to wisdom and knowledge of the will of God, Bacon uses God's word to prove that rhetoric and wisdom are separate. "The great difference between them," Bacon (1899, 176–177) suggests, "appears in the words of God to Moses upon his refusing…the charge assigned him: 'Aaron shall be thy speaker, and thou shalt be to him as God.' But for advantage and popular esteem, wisdom gives place to eloquence." Bacon, however, does not move as far away from rhetoric as Locke does and becomes an advocate for the teaching of rhetoric, mind you a very controlled and tranquil form of rhetoric.

Bacon (1899, 178) dispenses this advice concerning rhetoric: "by plainly painting virtue and goodness [rhetoric] renders, them as it were, conspicuous; for as they cannot be seen by the corporeal eye, the next degree is to have them set before us as lively as possible by the ornament of words and the strength of imagination." Rhetoric, plain rhetoric, still has value for Bacon because it helps build good character. Rhetoric becomes the means through and by which a concept or idea, previously ungraspable, can be seen more clearly. Rhetoric is an effective pedagogical tool as long as reason can maintain control over the "ornament of words" as Bacon refers to it. What is striking about Bacon's words, however, is that he sneaks in another concept that is rarely mentioned as important for the study of rhetoric and he introduces a fourth reason why rhetoric should be studied: rhetoric feeds and sparks the imagination. Prior to the passage cited above Bacon notes that rhetoric plays an important role in filling "the imagination with...observations and images as may assist reason. Rhetoric again is subordinated to reason in this passage but it plays a foundational role of feeding the mind, helping it to see more clearly, and to create more eloquently then one could without the aid of an imagination. Rhetoric now is important to study because it feeds the mind, it nurtures the soul, and invigorates creative thought.

This is also a theme Coleridge develops in his work. For Coleridge rhetoric is on the same level as reason. The imagination inspires and to know "the rules of the Imagination" is to know "the very powers of growth and production" (Richards 1965, 52). To master these rules, which have nothing to do with lists, manuals, or steps to follow, is to master reason and life. It is to discover beauty and the good life. "Further, and with particular reference to that undivided reason, neither merely speculative or merely practical, but both in one" Coleridge (Richards 1965, 52–53) announces, "I seem to myself to behold in the quiet objects on which I am gazing, more than arbitrary illustration, more than a mere simile, the work of my own fancy. I feel an awe, as if there were before my eyes the same power as that of the reason— the same power in a lower dignity, and therefore a symbol established in the truth of things...it becomes the visible organismus of the entire silent or elementary life of nature." Rhetoric in all its form whether speculative, practical, or a simile it is on the same footing with reason. It is harnessed in the power of the imagination which with the assistance of rhetoric taps into the life of nature or all that is true and good in the world. The imagination reconnects humans to nature. The imagination enables an individual to not only see the world of nature and the nature of the world but it tickles the "fancy" so one can see these worlds through their own mind's eye, in their own words, and own wonder. Rhetoric helps everyone to feel a sense of awe about the world and inspires one to create. It becomes a means to create, a mode of inquiry rather than a servant to reason and science.

It is Coleridge who inspires I.A Richards and Kenneth Burke to begin the process in the twentieth century to prepare for the return of rhetoric as a primary source of inquiry. Rhetoric, for Richards is a disciplined manner to correct misinformation. "The general task of Rhetoric," Richards (see Berthoff 1991, 94) makes clear in an essay on "The First Three Liberal Arts," "is to give...by exercise in comparisons, an insight into the different modes of speech and their exchanges and disguises...But we should do little good by explaining this, even with examples. All our pupils know it already. What they do not know is now to distinguish and meet the various modes of language in practice." Again we see the pedagogical importance of rhetoric. Rhetoric is the means by which students learn to distinguish between a good argument and one that is misguided. Students by nature, according to Richards, know instinctively when an argument sounds correct or not, but it is the role of the rhetoric teacher to create the skills within the students to understand why an argument is correct or not. Whereas traditional rhetoric is the art of persuasion, for Richards the new rhetoric is the art of stating a position. To state a position one needs to know basic structures of argument, sentences, and ideas. Once these basic structures are known then arguments can be evaluated. The art of stating a position also implies that an understanding of the context in which an argument is made. Richards (1936, 51) believes that "no words can be judged as to whether it is good or bad, correct or incorrect, beautiful or ugly, or anything else that matters to a writer, in isolation." As a result this system that Richards is proposing is not a structure that transcends time and therefore is universal. It is a historically bound approach to rhetoric that requires an understanding not only of the structures of language but also the context of the argument. Once the structure and context are understood then the heart of the "great and novel venture" of rhetoric is reached "the attempt to explain in detail how language works and with it to improve communication" (Richards 1936, 17–18). To understand what an economist is doing or to spark the imagination as it begins to go on a great venture, Richards attempts to restore rhetoric as a foundation for understanding what people do with language and how language is used to construct a world.

For Dalip Parameshwar Gaonkar, Kenneth Burke is the one person who invites rhetoric to return to the official realms of academia as an important element in understanding the imagination, the world, and academic inquiry. Burke releases the promiscuous nature of rhetoric into the world. He demonstrates how everyone uses rhetoric to police themselves thereby limiting one's imagination and ability to see life. Burke (1950, 35) in the *Rhetoric of Motives* argues "one can systematically extend the range of rhetoric, if one studies the persuasiveness of false or inadequate terms which may not be directly imposed upon us from without by some skillful speaker, but which we impose upon ourselves, in varying degrees of deliberateness and unawareness, through motives indeterminately self-protective and/or suicidal." With our self-discipline we limit the range in which rhetoric and by extension our minds can create. For Burke there is not necessarily some skilled manipulation going on externally by some word spinner or devious sort. A speaker is persuasive because the audience lacks the imagination to think differently from a prevailing mindset or discourse. Economic thought is one such example of how our self-discipline works. Burke turns to Bentham to show how this self-discipline might work through economics. "Bentham's utilitarian analysis of language," Burke (1950, 24) writes, "treating of the ways in which men find 'eulogistic coverings' for their 'material interests,' is thus seen to be essentially rhetorical, and to bear directly upon the motives of property as a rhetorical factor." Again we see an advocate of rhetoric warning against reducing its importance to economic exchanges, but for

Burke this is not a warning against smooth talking materialists who spout visions of glorious wealth and prosperity to the faithful. Burke is suggesting there is no need for these orators when people discipline themselves to reduce their own thoughts of the good life to any form of life that brings material goods and riches. As Bentham's witty phrase suggests we use rhetoric to justify greed. Rhetoric then becomes a means to search our own soul and flesh out those deceptive ways we justify actions that demean life and belittle the imagination. Rhetorical studies for Burke becomes an intense internal interrogation of the mind that can expose our eulogistic coverings of material desires. As Gaonkar points out Burke is the beginning of expanding the importance of rhetoric beyond the study of language. If rhetoric is used in economic exchanges then is it not possible that rhetoric is used in everyday language and in academic fields of knowledge beyond the traditional suspects such as speech? If speech is no longer the sole domain for rhetorical analysis then what about the written word? How does writing embody rhetorical strategies? (a question Rheinberger raised in his work in Chap. 1). And if the written word can also be interrogated and if rhetoric invades everyday communication then why should anyone believe that academic languages and modes of communication are free from rhetorical devices? Burke begins the process of turning rhetorical studies into a study of academic work. For the sciences Burke has unleashed the barbarians and since the 1990s the barbarians are looking everywhere in the academic and scholarly world to find rhetorical strategies.

3.2 Rhetoric as a Mode of Inquiry

The 1990s marks the return of rhetoric as an important field of study but instead of covering traditional areas such as speech and the classics, rhetoricians started looking at all academic fields of knowledge. Rhetoric, and this is the fifth reason why it is important for curriculum scholars to reclaim rhetoric, became a mode of inquiry, a method of investigation. As a leader in the return of rhetorical studies, Herbert Simon (1990, 6) described rhetoric as a mode of inquiry investigating all facets of academic work including "the rhetoric of the research report, the rhetoric of academic professions." All avenues of life are fair game. Rhetoric is no longer a tool experts use to judge the truthfulness of politicians, lawyers, and community leaders, now biologists, anthropologists, psychoanalysts, and even economists are the subjects of rhetorical analysis. There are some like Dalip Parameshwar Gaonkar who worry that rhetoric might be construed as a meta-discourse, serving as the arbiter of truth. Rhetoric as a mode of inquiry is not an attempt at adjudication to determine who is telling the truth and who is not, who is a trustworthy academic or economist and who is not. What rhetoric as a mode of inquiry seeks to determine how academics and scholars use rhetoric to construct an argument, a discourse, paradigm, a world view and how these rhetorical strategies shape the course of a field of knowledge or even a sub-discipline within a field of knowledge while eliminating alternative paths to knowing.

Rhetoric as a mode of inquiry is not only interested in how actors within a field of knowledge construct arguments, but also how rhetoric shapes how academics and scholars interact with outside forces as well. As John Lyne (1990, 37) writes rhetoric as a mode of inquiry helps everyone see "the values of scientific discourse negotiate with state ideologues, or with private interests, different rhetorical strategies bring these different vectors to bear in different ways." Lyne's statement in itself justifies the use of rhetoric to study economic thought. Economists are more likely than most academic fields of knowledge to interact with state institutions and private interests. When it comes to economic thought and rhetoric as a mode of inquiry the most pressing question is how economic thought is interpreted and utilized by state powers and private interests to justify certain policies and individual practices to protect wealth and nurture an imbalance of power in society. It is also important to find out how neoclassical ideas became codified and fossilized as the only legitimate mode of thinking in state institutions and private interests and why alternative modes of thinking are ignored and rarely entertained as legitimate substitutes for neoclassical economics.

Since the interactions between a field of knowledge and state and private institutions are a concern for those who approach rhetorical studies as a mode of inquiry a major shift in attention occurs. While the individual is still of concern, what becomes a primary interest is the way in which rhetorical strategies shape discursive communities. Rhetoric as a mode of inquiry, Lyne (1990, 47) believes "configures activity within a field so that one can picture the subject matter not just as a state of affairs but as a direction of affairs." The rhetoric members of a field of knowledge use not only shapes the direction of research questions, professional standards, ways in which phenomenon can be addressed but also limits how one can legitimately think about an issue without being labeled a "heretic" and denied access to a field of knowledge. The tighter members of a discursive community guard their boundaries the less likely alternative modes of inquiry and alternative paths to knowledge will be tolerated. In regards to rhetoric as a mode of inquiry the art of persuasion is still an important issue to decipher but it is no longer necessarily the most articulate person or knowledge community that wins the day and defines the research agenda. Now it is more about who is most persuasive in defining the direction research should take and what questions are worth addressing and by extension which questions or topics are not considered legitimate.

Dalip Parameshwar Gaonkar considers the 1990s as a watershed mark since rhetoric is no longer seen as merely an exercise of ornamental excess. Rhetoric as a mode of inquiry is not as Gaonkar (1990, 343) notes something to be feared because it can "spring up at any time, from within or without, to pollute and possess what is not its own for the sake of temporary advantage and gratification. Thus rhetoric is seen as a nomadic discipline that threatens the integrity of the republic of knowledge." Now, rhetoric as a mode of inquiry is the integrity of a field of knowledge. It is not a pollutant that if left unguarded can destroy academic boundaries, instead rhetoric is the academic boundaries, it is the air scholars' breath to shape policy, define research agendas, adopt theories, and mold the course of intellectual development. Again Gaonkar (1990, 349–350), this time calling on the support of Kenneth

Burke, suggests that "'by showing how a rhetorical motive is often present where it is not usually recognized, or thought to belong'" rhetoric as a mode of inquiry demonstrates how acts of persuasion are embedded in all academic discourse and not alien to these discourses but naturally present.

Gaonkar warns that natural presence of rhetoric in all academic discourses presents a special problem for rhetoricians of knowledge. To suggest rhetoric is embedded in academic work then there is the risk of mistaking this reality with the belief that rhetoric is the monarch, patriarch, and god of all academic knowledge. This proclamation can take the shape of what Gaonkar (1997) calls implicit analysis or the idea that a user of rhetorical strategies within a field of knowledge is unknowingly using rhetoric to construct an argument and is in need of a "special reader" of this rhetoric to become aware how it is being used. In a sense there is no way to completely avoid the label of "implicit reader" especially if one is perceived as an outsider. There are rhetoricians of specific fields of knowledge who avoid this label because they come from within the field. Deirdre McCloskey for instance was successful in demonstrating how economics is rhetorical because she was trained in neoclassical economics and already established in the field. The same is true for Arjo Klamer and Philip Mirowski they are trained economists. Robert Solow when he turned to rhetorical analysis was not only a trained economist but a Noble Prize winner which meant his "voice of reason" was more amplified and more likely to be listened to. However, if curriculum theorists and other scholars are to try to change the course of economic thought and policy formation, then we are going to have to simultaneously avoid the "special reader" label while bringing something important to the discourse of neoclassical economics and entering into a conversation with economists risking being ignored because we lack the assumed necessary credentials to speak in the correct voice and manner.

Wayne Booth is another rhetorician who views rhetoric as a mode of inquiry. Like Gaonkar, Booth (2004, 8) challenges the "merely crowd" and dismisses their dismissiveness when he argues "all of us [rhetoricians] view rhetoric as not reducible to the mere cosmetics of real truth or solid argument: it can in itself be a mode of genuine inquiry." Rhetoric as a mode of inquiry is essential for democratic engagement. Booth pulls out the tradition card and suggests that rhetoric as a mode of inquiry creates "the character, the ethos, of those who engage with it. This is why the quality of our citizenry depends on whether their education has concentrated on the productive forms of rhetorical engagement" (Booth 2004, 18). Booth's argument is also a reason why neoclassical thought needs to be engaged rhetorically. Politicians and corporate leaders, with no real resistance from economists, have condensed democratic life to a simple but dangerous equation: democracy equals free markets and consumption of goods. This often unchallenged assumption of what democracy is answers Booth's query about the education of citizens. Citizens have been educated to accept this crude equation and as a result the democratic foundations in the United States are very weak. Democratic freedoms have become attached to the vicissitudes of the stock market or the sermons of Milton Friedman or Arthur Laffer who threaten doom if any of the holy principles of supply side economics are tinkered with.

In response to the challenges to a vibrant democracy, Booth (2004, 46) recommends the creation of a "listening rhetoric." Listening rhetoric…is what I most long to celebrate and practice—the kind that is sadly rare. At its deepest levels it deserves my coinage; rhetorology—an even deeper prober for common ground." This common ground does not appear magically with little effort at critical dialogue and it never appears when the participants in a dialogue are on unequal ground such as when a doctor tells a patient to do something and the patient does not probe as to the effects of the advice but merely says "yes doctor." How can a listening rhetoric be created that simultaneously looks at economic rhetoric critically especially neoclassical claims of truthfulness that are assumed to be natural economic conditions and at the same time recognize the importance economics plays in everyday decisions?

3.3   The Rhetoric of Science

One way of addressing this question of creating a critical dialogue with economists has already been addressed. It is important not to utilize rhetoric as a mode of inquiry in order to act as the "special reader" who will reveal the code neoclassical economists unwittingly use. Even if this approach would be attempted it will fail because in the power game called academic discourse and politics it is clearly than probably anything in academic work that economists have most of the power on their side. The only rhetorical and political power curriculum scholars have is a vibrant concept of a living, dynamic democracy in which economics plays a lesser role and other definitions of freedom regain influence and power.

There are other ways not to approach a dialogue between neoclassical economics and curriculum scholars. Rhetoric as a mode of inquiry is not a Gotcha! Game. Rhetoricians are not out to show that scientists play rhetorical games. There are scientists who still live in the seventeenth or eighteenth century and believe that rhetoric is just as Locke said: unnecessary ornaments that deceive. This is just a rhetorical ploy denying that rhetoric is used while rhetorical devices are being used. If the task of rhetoric was to just show how ploys are deployed through language there would be no need to focus on the sciences. Rhetoricians could continue to focus, as Aristotle advised, on lawmakers, civic leaders, and eulogists to demonstrate how rhetoric works. Nor is it wise to deploy rhetoric as a mode of inquiry in order to do what Robert Solow (Klamer and Leonard 1996, 21) refers to as the "look, Ma a metaphor stage" in which readers are astonished to discover that scientists are humans too and actually do use language in ways mere non-scientists do. Scientists like anyone else are steeped in rhetoric and this is one reason why it is interesting to study the rhetoric of economics. Scientists like those in any other culture use rhetoric in specific ways in order to invent a world, inquiry about the world, and construct ways to understand the world. To write home to mother in astonishment when you discover that scientists do indeed function within a culture is to fall right into the scientism camp that is a prime barrier to disciplinary border crossing. "Scientism," Philip Mirowski (2004, 64) suggests, "here denotes the overweening confidence and chauvinism on the part of those inducted into a natural science that their own local culture represents everything noble, rational, and efficacious about the human race; and further, anything less than abject tribute and total capitulation to this position…on the part of those without passports from the culture in question should be met with scorn, ridicule, and contempt." Why would we as curriculum scholars make our task more difficult since we do not have the proper passport and act in amazement when it is discovered that scientists in general and economists specifically are deft at deploying rhetorical strategies?

Scientism is a mask some still hide behind but to use rhetoric as a mode of inquiry to suggest that because of the mask of scientism, scientists and economists are acting maliciously will not work as a strategy either. Instead I think we should take Nietzsche's (1974, 121–122) advice he dispenses in an aphorism titled "Only as Creators!" in *The Gay Science*. I cite the passage at length in chapter nine so I will only summarize here. For Nietzsche only creators of knowledge can challenge another system of knowledge. Creators destroy established systems and in the process of creating new names and "new things" a new way of thinking is presented. This Nietzsche believed was enough to usher in a new paradigm or epistemic shift. Rhetoric as a mode of inquiry becomes a creator, a new, alternative way to think about economics and to ask different questions or the same questions in different ways in order to create new things. What those new things can be will be determined by what is destroyed but what is destroyed can be done only by a creator of a new system or a new way of looking at an old system. For instance, why do neoclassical economists assume that economic actors are rational and why do they assume they only act in selfish ways? Why can't Smith's baker not make bread for other reasons than his/her own selfish needs? Why does Friedman's maxim to maximize profits have to be the only focus of a corporation? To destroy the limits of Smith, Friedman and other economists, we have to create. To create is not to expose the rhetoric of neoclassical economists but to utilize rhetoric as a mode of inquiry to undress the skin of the neoclassical way of thinking in order to create new names, estimations or possibilities. This is enough to do in order to create a new world and a new way of thinking.

Nietzsche's aphorism fits perfectly with Alan Gross' (1996, 3) views concerning the rhetoric of science. Science like any other field is primarily "the creation of knowledge" and the creation of knowledge "is a task beginning with self-persuasion and ending with the persuasion of others." The task of the scientist then is to convince oneself first that what has been invented is the truth or a representation of nature and then to convince others of this condition. It is the task of the rhetorician to convince the scientist that this process of self and community persuasion is a very complicated, often truthful, and potentially transformative, but always inventive. In between this process of self-persuasion and the persuasion of others is a whole gambit of people, objects, other non-human animals, rhetorical devices, and other cultural artifacts that have to be accounted for. There is the laboratory, if a laboratory is needed, the technicians and scientists of the laboratory, the funding sources and the cultural exchanges between the scientists and the holders of the purse strings, the machines to invent the illusion that the laboratory indeed is nature, animals to test the hunches scientists follow, journals and publishers fully equipped with editors and review boards, ethic committees, internal review boards, university administrators, students, corporate heads, venture capitalists, and on and on the list could grow.

Gaonkar suggests that the rhetoric of science has stalled. "The Fact is," Gaonkar (1997, 41) diplomatically and rhetorically posits, "that RS [rhetoric of science], like so many other research projects based on a revived interest in rhetoric, has stalled after a promising beginning. There are two possible explanations as to why the RS project has stalled: first, it has misread science; second, it has misappropriated… rhetoric." Have I missed the boat? Am I 15 years too late in my thinking? These are rhetorical questions of course. I have not. Gaonkar's suggestion is when it comes to philosophical and historical questions concerning science the rhetoric of science is not equipped to respond in any fruitful way. Let the philosophers and historians handle the tough foundational questions about science and the rhetoricians can show the inner workings of science through case studies. Gaonkar is afraid to cross disciplinary boundaries and does not want to tread on the feet of philosophers and historians of science, therefore rhetoricians should know their limits or they will fail. It is in the rhetoric of economics that rhetoricians are exposing Gaonkar's overly cautious approach. Gaonkar in his thinking has constructed concrete barriers between philosophy, history, and rhetoric. While these three fields are different, their intellectual boundaries can be crossed and the rhetoricians of economics are leading the way. In fact, they have already crossed over the borders numerous times. It is with this border crossing in mind that I will now turn to the rhetoric of economics.

3.4   The Rhetoric of Economics

Throughout this chapter I cited phrases and aphorisms rhetors of the past used to warn against the use of rhetoric for private gain or public manipulation. Most of those warnings contained some reference to economics or reduction of life to economic exchanges. I did this to make the point that economics has always been rhetorical. Perhaps not in the most flattering of light, but in the forefront of the minds of oratory masters was the role economics plays in developing public leaders and philosophers. Adam Smith was a professor of rhetoric and moral philosophy and it is clear the role these disciplines played in his thinking about economic thought. Smith's lectures were re-discovered and in 1958 published in a book form as *Lectures on Rhetoric and Belles Lettres*. Smith's approach to rhetoric was a combination of a classical Aristotelian and typical eighteenth century approach to the subject as Wayne Booth (2004, 29) notes he "saw it as serving other ends than the search for truth, even while being indispensable. One can see an influence of rhetoric in the thinking of other classical economists such as Jeremy Bentham as noted above and in John Stuart Mill who is viewed as an early advocate of neoclassical economic thought with his focus on utility. As neoclassical economics rose to dominance by the late nineteenth century and early twentieth century, rhetoric remained influential but it was not viewed as an essential or important dimension of economic thought to recognize or comment on. This state of affairs has changed since Deirdre McCloskey's work in 1985.

McCloskey's work first appeared in the *Journal of Economic Literature* in 1983 and soon after that in 1985 as a full length book, *The Rhetoric of Economics*. It is McCloskey who returned economists, whether they like it or not, back to their rhetorical roots. For McCloskey economists use rhetoric the same way others do in other scientific fields and in the humanities. Economists deploy rhetoric to persuade themselves and others, educate, (mis)inform, indoctrinate, and even deceive. This is not anything revelatory or new for any academic discipline, what McCloskey's work did was force economists to admit how central rhetoric is to economic thought. For McCloskey the purpose for studying the rhetoric of economists is not to embarrass economists or knock them off their privileged and "scientific" perch but to understand what it is economist do when they construct models, develop truths, and persuade others to think in a specific disciplined and systematic manner as opposed to other disciplined and systematic ways of thinking.

The purpose of rhetoric is to think. This is McCloskey's goal in understanding how economists use rhetoric. The "unexamined metaphor," McCloskey (1998, 46) "is a substitute for thinking—which is a recommendation to examine the metaphors, not to attempt the impossible by banishing them." The point of understanding the rhetoric of economics is not to seek out a full proof system to avoid rhetoric, this system does not and will never exist, but the rhetoric of economics is an invitation for economists, and I would add non-economists, to think about what it is they do when they think about economics and invent economic theories. McCloskey gets even closer to the nerve center of academic disciplines such as economics and suggests that the purpose of the rhetoric of economics is to ask how economists establish their authority. "Ethos, the Greek word simply for 'character'," McCloskey (1998, 7) submits, "is the fictional character an author assumes. It is the same as the Latin persona or the modern 'implied author.' No one can refrain from assuming a character, good or bad. The exordium, or beginning of any speech must establish an ethos worth believing." How have economists become such trustworthy character witnesses as to what is going on in a complex, unruly, uncontrollable, chaotic economic system called capitalism? How did economists become the high priests of what is right and good in a "free market" economy? How did economists become so good at their craft that so few people question their assumptions, goals, and conclusions? Rhetoric, of course. But it is not a rhetoric of deception. It is a rhetoric of authority and persuasion that can at times be deceptive and is always incomplete.

There are numerous reasons why the authority of neoclassical economics emerged as the only alternative to thinking about economics. Those who eventually became early neoclassical economists like Vilfredo Pareto, Leon Walrus, William Jevons, and Carl Menger were able to latch their ideas onto two very important nineteenth century symbols of academic truth, physics and mathematics, and ride the authority of these fields to promote their own field of study as a science. When

I discuss Philip Mirowski's ideas concerning the rhetoric of economics later in this chapter I will discuss the role of physics and mathematics more. In the twentieth century neoclassical economists were able to claim authority by discrediting alternative views which continues to this day with many neoclassical economists still hell bent on discrediting any form of Keynesian economics. To banish alternatives thought it took more than just ideas. It took money; and neoclassical economists at the University of Chicago and the Mont Pèlerin society were always well funded just as they are today through the Heritage Foundation and the Cato Institute which is to say through the Koch Brothers. This is a topic that will be discussed later as well. The role of money does suggest that the authority of neoclassical economists was bought and there is a case to be made, but as McCloskey notes authority can never be solely bought it has to be earned through persuasion.

In his study McCloskey also took on the deniers, the anti-rhetoric advocates who rolled out the traditional scientism weaponry of objectivity, value neutrality, apolitical commitments, and of course, clear and concise and therefore anti-rhetorical prose. To the converted, which is to say the blinded, these principles all sound good and sufficient to guard against any human taint to the pursuit of "Truth". While guarding against subjectivity and personal values and avoiding any political entanglements that will skew any research before it is developed is the prudent course to take, objectivity, value neutrality, and apolitical commitments are all arts not hard and true values. There is an art to become as objective and value neutral as possible and there is a skill to navigate all the political dimensions of doing science. But it is impossible to avoid subjective beliefs, values, and politics. Neoclassical economists utilized physics and mathematics in ways that puzzled physicists and mathematicians, gravitated towards the values espoused by Friedrich Hayek and now Milton Friedman and Gary Becker, and are so entangled in politics it is sometimes hard to know where economics begins and political ideology ends. And yet, neoclassical economic thought is still king in the economic and political arenas. By their own standards of objectivity, value neutrality, and apolitical commitments neoclassical thought should be excluded from any consideration when discussing these principles. McCloskey in his ground breaking work demonstrates how this anti-rhetoric take on economics just was not realistic or tenable.

Like all good revolutions, it was not long before the children of the revolution were not satisfied with the first generation. Arjo Klamer and Philip Mirowski both believe that McCloskey did not go far enough in her analysis of the rhetoric of economics. Klamer is best viewed as a bridge. As a doctoral student and young scholar he certainly was involved in the early days of bringing rhetoric back to the forefront of economic thinking. Nonetheless he tried to push rhetorical analysis further than McCloskey's work did. Klamer wishes to expound more on how a metaphor works in economics. As Klamer and Leonard (1996, 21) notes critics of McCloskey responded to her work and conclude that "'If metaphor occurs in economics so what?—its existence is incidental to the business of doing economics.'... A more useful inquiry...will examine how metaphors actually work in economics." This is what Klamer set out to do. In fact before I delve into Klamer's look into economic metaphors I want to show how metaphor is the usual and general topic discussed when covering rhetoric in general and the rhetoric of economics specifically.

In his Flexner Lectures to the Bryn Mawrters in the audience in 1935, I.A. Richards (1936, 90) said "[t]hroughout the history of Rhetoric, metaphor has been treated as a sort of happy extra trick with words, an opportunity to exploit the accidents of their versatility, something in place occasionally but requiring unusual skill and caution." Metaphor, like rhetoric in general, is treated with general suspicion. This suspicion makes one wonder how something so fundamental in life such as metaphor and rhetoric can be viewed with such disdain and skepticism. If rhetoric and by extension metaphor is in part what makes humans human why has there been such great distrust and such great effort to distant oneself from an important component of humanity? Richards countered this view of metaphor with his own definition of the importance of metaphor. "That metaphor is the omnipresent principle of language," Richards (1936, 92) asserted, "can be shown by mere observation. We cannot get through three sentences of ordinary fluid discourse without it... Even in rigid language of the settled sciences we do not eliminate or prevent it without great difficulty....our constant chief difficulty is to discover how we are using it and how our supposedly fixed words are shifting their senses." Richards's metaphor endows our language with a rhythmic flow even in the most rigid of human endeavors of truthfulness like the sciences and demonstrates how this fluidness acts like a river and reshapes itself through the senses. Metaphor is everywhere, even in economics, the task of human beings is to constantly figure out how metaphors are being used and how they are changing our sense of reality.

The fluid metaphor Richards (1936, 94) adopts demonstrates that by its nature metaphor is "a borrowing between and intercourse of thoughts, a transaction between contexts." Metaphor by its nature is an economical transaction that acts like a currency, exchanging hands and contexts as meaning (value) shifts depending on the context. Just as currency is a prime example so is metaphor of what J. Hillis Miller following Austin and Derrida calls iterability or the ability to take a word or concept and transplant it from one context to another and by this transaction either maintaining the original intent of the word or concept or creating a new meaning for it. The key for Richards is to have a command of metaphor. Command of metaphor is the key to a good life. "Words are not a medium in which to copy life. Their true work is to restore life itself to order" (Richards 1936, 134). Moreover, Richards (1936, 135) continues: "A command of metaphors—a command of the interpretation of metaphors—can go deeper still into the control of the world that we make for ourselves to live in." This is exactly what economists have done and one reason why neoclassical economists enjoy such strong favor and statue in our society. They have been able to use metaphors to shape what the good life may mean, even if in reality it is just life, and through the command of metaphor they are able to shape the world and its potential meaning.

Nietzsche also saw the importance of metaphor in shaping reality. Nietzsche asks (Derrida 1982, 217):

> What then is truth? A mobile army of metaphors, metonymics, anthropomorphisms: in short, a sum of human relations which become poetically and rhetorically intensified, metamorphosed, adorned, and after long usage, seem to a nation fixed, canonical and binding; truths are illusions of which one has forgotten that they are illusions; worn out metaphors

which have become powerless to affect the senses,...coins which have their obverse... effaced and now are no longer of account as coins but merely as metal.

We see in Nietzsche the same use of metaphor as in Richards. Truth is an economic transaction, a coin, that has forgot how many times it has been passed around as a coin and as a results sees itself as a truthful, fixed reality rather than part of an illusory mobile army.

This aphorism from Nietzsche is pulled from Derrida's work "White Mythology." Derrida like Nietzsche and Richards joins the fray. Derrida adopts similar metaphors to describe the nature of the metaphor. He argues that the use of metaphor is not an accident but is something humans do subconsciously and naturally. To make his point he relies on a common metaphor, usury and then the coin. Both economic metaphors used to describe something that is done every day. A metaphor is not only a poetic device that critics of rhetoric like Locke and empiricists from the nineteenth century disdain, but as Derrida, Nietzsche, and Richards demonstrate metaphor is embedded in economic exchange and economic thought. Metaphor adds, creates, invents, imagines, and transfers. "And I have been able," Derrida (1982, 213) submits "without sacrificing fidelity, to substitute one for the other." Derrida, like anyone else is able to create a world, invent meaning, and transfer concepts from one context to the next through the use of metaphor without demeaning truth or reality. Citing Aristotle Derrida (1982, 231) writes that "'Metaphor...consists in giving... the thing a name...that belongs to something else...the transference being." This is exactly what economists of all sorts have done before the invention of their field of knowledge. Before the field of economics existed, all people were economists; transferring, transacting, and transforming their lives. Professional economists have just continued this tradition.

We can see this process take place when the General Theory of Equilibrium is discussed. One of its inventors, Kenneth Arrow, has warned that this economic concept of perfect competition is not as universal as neoliberals fancifully dream. For Arrow there is at least one area that does not fit the formula too well: health care. However, what is occurring as I write is an attempt by conservative politicians to force health care into a neoliberal political ideology with neoclassical economists as their major rhetorical weapon. The consequences have not been so healthy but the ideological points scored seem to please conservative politicians who despise universal, state controlled health care more than they seem to dislike communism or Karl Marx. How Strange a brew USA conservatives are fueld by hysterics and ressentiment.

This is where Arjo Klamer and Thomas Leonard enter into our story on metaphor. Metaphor is fine for poets to play with but it is "anathema for scientists" because metaphors "introduce ambiguity" (Klamer and Leonard 1996, 20). So why are metaphors so important to economics but economists are so cool on their presence? To put it another way why are economists so embarrassed to be associated with a concept that is so vital to the development of their field? The truth is they are not. Economists like anyone else like to play with metaphors and understand that if they craftily construct a metaphor it will only enhance their authority as speakers of economic truths. "A scientific metaphor" may be ambiguous but it "is propositional; it only invites further inquiry. It does not presuppose or by itself settle the similarities between the principal and subsidiary subjects. The task of interpretation remains. It is this open-endedness and lack of explicitness that makes metaphor so useful to scientific inquiry" (Klamer and Leonard 1996, 30). It is the need for interpretation and the ambiguity as to what a metaphor or economic concept may mean that creates the need for someone to supply meaning. The person who does so will be viewed as an expert, therefore, through the use of a metaphor economists are merely taking advantage of an opportunity to enhance their position within society as someone whose advice needs to be sought.

For Klamer and Leonard there are different ways economists use metaphors in order to enhance their authority. First there is the pedagogical metaphor. "Effective pedagogical metaphors typically provide mental images…with which the audience can visualize an otherwise complicated concept" (Klamer and Leonard 1996, 31). An example of a pedagogical metaphor is the currently popular entrepreneur as innovator. The Metaphor of innovator constructs an image of the entrepreneur as a catalyst for change, who, it is implied, will always be for the common good and serves as a better descriptive purpose than other terms such as venture capitalist who often connotes opportunism and recklessness. In contrast the entrepreneur as innovator is a benevolent soul, often seen as taking all the risk and sharing the benefits if successful. It is an easy metaphor to relate to because anyone can see themselves as the benevolent giver and it is a metaphor that is retroactive. The economist does not have to rush to use this pedagogical device in the present or future except in the abstract. The past is where the entrepreneur as innovator is applied and the past always supplies examples. It is now common and easy to label Mark Zuckerberg as an innovator because everyone knows the success of Facebook and when the movie *The Social Network* appeared, and Zuckerberg's fame already established, it was easy (natural?) to view the Winklevoss brothers as spoiled, envious rich kids looking to leech off of the successful entrepreneur because the pedagogical metaphor had already taught us who is the benevolent one and who was to play the villain.

A more important metaphor is the heuristic metaphor. A Heuristic metaphor "serves to catalyze our thinking, helping to approach a phenomenon in a novel way….Metaphor is cognitive here because its respective subjects interact to create new meaning" (Klamer and Leonard 1996, 32–33). For me the classic heuristic metaphor is Adam Smith's invisible hand. *The Wealth of Nations* attempts to introduce a novel way of thinking about economic interactions and challenging established modes of thinking about economics, politics, and social order. When one is challenging the accepted order of things it is not those with invested power who question themselves and wonder if the challenger is maybe right. The challenger is doubted and questioned. Smith used the metaphor of the invisible hand as a device to preempt questions concerning the rise of conflicts in a "free" market economy. This heuristic device acted as a way for Smith to deflect criticisms of his proposed system and to create enough space to address all the unknown problems that would surely emerge if the old economic order were to be uprooted and a new way of doing business were to be implemented. The invisible hand could be called upon to serve as the future arbiter of problems and help new converts to "free" market capitalism to visualize such a system working in real life situations.

A final metaphor Klamer and Leonard discuss is what they refer to as a constitutive metaphor. "These metaphors work on an even more fundamental level. Constitutive metaphors are those necessary conceptual schemes through which we interpret a world that is either unknowable…or at least unknown….When we say that a metaphor <frames our thinking>, we mean to say that such metaphors profoundly influence our thinking, what we see and hear" (Klamer and Leonard 1996, 39–40). Economic models have come to serve as constitutive metaphors. The world of economic exchanges is too complex to understand or predict what is going on and what might happen. Instead of dealing with the unknowable economists rely on models that reduce the unruly world to more manageable bites of information so that some mechanism of prediction and reliability can be used to generate economic laws and develop policies. These models, however, in no way reflect reality. They frame our thinking and shape "what we see and hear." Another example is the early neoclassical economists' use of physics to privilege utility as the most important concept in economics. As Philip Mirowski noted the process of converting the physics notion of energy to an economic principle of utility is fraught with error and misplaced principles. Yet, the conversion was successful. Neoclassical economists used a nineteenth century theory of energy called energetics introduced by the Scottish physicist William Rankine whose cause was later taken up by Georg Helm and Friedrich Wilhelm Ostwald. Energetics was seen as the answers to the physicists' questions concerning energy. In physics this claim turned out to be false but in economics energetics became the catalyst to create a mathematical understanding of utility. While physicists and mathematicians were baffled by what economists were doing with energetics and mathematics, their opinions did not matter. Neoclassical economists were able to create a general and universal concept of utility. The concept of utility continues to shape how neoclassical economists see and think about economics. As Klamer and Leonard (1996, 43) note constitutive metaphors "are not picked up and discarded like heuristic metaphors or mere preferences; constitutive metaphors are us. A fundamentally changed perspective… requires changing oneself." And the neoclassic age of economics began, deep in the bowels of the ambiguous and uncertain world of the metaphor and rhetoric. Now, the denial game that rhetoric had ever been used began too.

3.5 Mirowskian Rhetoric

Often when a new way of thinking about or living in the world is invented a founder is named and the disciples labeled. When Newton co-created calculus and explained planetary motion, Newtonians soon were created. When Marx's ideas were adopted throughout the world, Marxists were born. The same can be said about the rhetoric of economics. After Deirdre McCloskey's ideas took hold the McCloskians were born and the greatest of these is Philip Mirowski. Note that when disciples follow a

new intellectual or political movement, they are not expected to be faithful followers nor are they expected to receive the endorsement of the leader. Although I do not
know of any criticism McCloskey has levied against Mirowski, I do know that
Mirowski has extended McCloskey's ideas further than anyone else and his work is
the best in the field.

Like any good revolutionary Mirowski did not believe that McCloskey went far
enough to explain the role rhetoric played in the shaping of economics. "Hence,
rhetoric," Mirowski (1988, 118) submits, "is also a form of a theory of social order,
a prototype of morality, statecraft, and of philosophy itself." With this in mind,
Mirowski's work takes a look back to the nineteenth century and the rise of neoclassical economics and the use of an outdated and discredited physics concept of
energy to create a mathematical concept of utility. The use of physics and mathematics should be viewed as an attempt to build a defense around utility to prevent
any attempt to discredit it as a force in economic thought. If the attackers were
physicists or mathematicians it did not matter because in the rhetorical struggles of
knowledge building they did not matter. The use of physics and mathematics was to
defend the concept of utility from any criticism from economists. "The Physics
metaphor [of Energy in general and Energetics specifically] implies that economics
is a science and deserves all the legitimacy that is granted to physics itself because
there exists no great difference between the two modes of inquiry" (Mirowski 1988,
141). No one within the realm of economics in the nineteenth century would dare to
question the "Holy trinity" of science, physics, and mathematics. If they did they
risked their own reputations and more importantly exile from the field of economics. The use of physics and mathematics in the name of science worked for the early
neoclassical economists. Now their early rhetorical work holds just as Nietzsche
warned it would. It has been forgotten as rhetoric and now lives the illusion of truth.
As Mirowski (1988, 130) noted "the early neoclassicals took the model of "energy"
from physics, changed the names of the variables, postulated that 'utility' acted like
energy, and then flogged the package wholesale as economics."

There is yet another side of this story, however. It is not just about neoclassical
economists adopting a physics concept of energy to fit their mathematical and theoretical needs, but economists were able to gain legitimacy as a science because scientists adopted economic metaphors. For instance, Mirowski (1989, 125) writes
"that the first law of thermodynamics is derived from what was essentially an economic metaphor" because of the "French quest for a quantification of work, and the
fact that Carnot himself had been a student of early-nineteenth –century French
economics." In another case dealing with Hermann von Hemlholtz, Mirowski
(1989, 131) demonstrates how economic thought infiltrated the thinking of physicists. "For Helmholtz, the world is a machine; man is a machine. Men work for men;
the world is a machine; you can't swindle Nature; in the natural state no one is
swindled." As the notion of energy changed so did the notion of value. Instead of
holding firm to the idea that money can only grow as long as it had something of

value supporting it, a more abstract notion of money emerged at the same time that physicists abandoned a notion of energy as the only force of importance. This "breakdown of energy" as the force necessary for sustainability in science, economics, and life for Mirowski (1989, 137) reveals underneath the surface of theoretical physics, neoclassical economics, and other disciplines a reality based in the notion that "Metaphors of the body, of motion, and of value are more perfectly reconciled by means of the realization that each is a fiction, but the same fiction, a fiction necessary for the organization of human discourse."

On one level, neoclassical economists need to be applauded for their adeptness in playing academic politics and winning a major seat not only at the tables of every university but a major chair in all boardrooms and political capitols. They were able to maneuver themselves into powerful positions because of their rhetorical skills, and of course later on, their denial of ever using their rhetorical skills. "I would insist," Mirowski (2004, 34) posits, "that the fundamental strength of the neoclassical model derives…from its attempts to embody popular images of scientific inquiry and content into what purports to be a description of social interaction." Neoclassical economists have presented a vision of how human beings interact in the world, and they have been skillful and disciplined in developing this vision by turning it into a science. But there has been a price, not just a monetary economic one either, for the success of neoclassical economics. It has done exactly what anti-rhetoric advocates like economists have argued when warning about the dangers of rhetoric. They have, and more importantly they have allowed political ideologues, to overstate their claims. They seek to create a knowledge monopoly and in many ways they have succeeded already in the academic and political realm. As Mirowski (2004, 378) points out "[b]y bequeathing to people a stark yet seemingly paradoxical image of what they want, neoclassical economics can get what it wants, which is to subsume or displace all other academic social theory in the name of unified science." By pretending to understand the needs and desires of individuals, reducing these needs to a few economic concepts, neoclassical economics is attempting to do what philosophers, theologians, physicists, mathematicians, and poets all failed to do one time or another in intellectual history: monopolize thought in the name of science. This power play can only fail, but what will the cost of failure be? We can already see some of the consequences of neoclassical power plays. Life is reduced to economics, a university education is defined by job security, sports, like other forms of entertainment, are defined as businesses, and the only "reality" that seems to matter in all realms of life is the so-called "economic reality." As a result of the meta-fiction called neoclassical economics theology, history, the fine arts, philosophy, literature, and the sciences have all suffered from a myopia that limits creativity, imagination, and future possibilities for life to flourish. It is scholars such as Mirowski and his creative and rhetorical abilities that can help everyone escape the neoclassical trap and redefine life.

References

Augustine. (2008). *Confessions*. Oxford: Oxford University Press.
Bacon, F. (1899). *Advancement of learning and Novum Organum*. New York: The Colonial Press.
Berthoff, A. (Ed.). (1991). *Richards on Rhetoric: I.A. Richards selected essays, 1929–1974*. Oxford: Oxford University Press.
Booth, W. (2004). *The rhetoric of rhetoric: The quest for effective communications*. Malden: Blackwell Publishing.
Brickhouse, R. (2010). *The puzzle of modern economics: Science or ideology?* Cambridge: Cambridge University Press.
Burke, K. (1950). *The rhetoric of motives*. New York: Prentice-Hall.
Derrida, J. (1982). *Margins of philosophy* (trans: Bass, A.). Chicago: University of Chicago Press.
Folbre, N., & Hartmann, H. (1988). The rhetoric of self-interest: Ideology of gender in economic theory. In A. Klamer, D. McCloskey, & R. Solow (Eds.), *The consequences of economic rhetoric* (pp. 184–203). Cambridge: Cambridge University Press.
Gaonker, D. P. (1990). Rhetoric and its double: Reflections on the rhetorical turn in the human sciences. In H. Simons (Ed.), *The rhetorical turn: Invention and persuasion in the conduct of inquiry* (pp. 341–366). Chicago: University of Chicago Press.
Gaonker, D. P. (1997). The idea of rhetoric in the rhetoric of science. In H. Simons (Ed.), *Rhetorical hermeneutics: Invention and interpretation in the age of science* (pp. 25–85). Albany: SUNY Press.
Garver, E. (1994). *Aristotle's rhetoric: An art of character*. Chicago: University of Chicago Press.
Gross, A. (1996). *The rhetoric of science*. Cambridge, MA: Harvard University Press.
Klamer, A., & Leonard, T. (1996). So what's an economic metaphor. In P. Mirowski (Ed.), *Natural Images in economic thought: "Markets read in tooth and claw"* (pp. 20–51). Cambridge, UK: Cambridge University Press.
Lyne, J. (1990). Bio-politics: Moralizing the life sciences. In H. Simons (Ed.), *The rhetorical turn: Invention and persuasion in the conduct of inquiry* (pp. 35–57). Chicago: University of Chicago Press.
McCloskey, D. (1985/1998). *The rhetoric of economics*. Madison: University of Wisconsin Press.
Mirowski, P. (1988). Shall I compare thee to a Minkowski-Ricardo-Leontief-Metzler matrix of the Mosak-Hicks type? Or, rhetoric, mathematics, and the nature of neoclassical economic theory. In A. Klamer, D. McCloskey, & R. Solow (Eds.), *The consequences of economic rhetoric* (pp. 117–145). Cambridge: Cambridge University Press.
Mirowski, P. (1989). *More heat than light: Economics as social physics, physics as nature's economics*. Cambridge, UK: Cambridge University Press.
Mirowski, P. (2004). *The effortless economy of science*. Durham: Duke University Press.
Nietzsche, F. (1974). *The gay science*. (Walter, Kaufmann, Trans.). New York: Vintage Press.
Richards, I. A. (1936). *The philosophy of rhetoric*. Oxford: Oxford University Press.
Richards, I. A. (1965). *Coleridge on imagination*. Bloomington: Indiana University Press.
Simons, Herbert. 1990. Introduction: The rhetoric of inquiry as an intellectual movement. Herbert Simons, The rhetorical turn: Invention and persuasion in the conduct of inquiry, 1–31. Chicago: University of Chicago Press.
Smelser, N. (Ed.). (1973). *Karl Marx: On society and social change*. Chicago: University of Chicago Press.
Sperber, J. (2013). *Karl Marx: A nineteenth-century life*. New York: Liveright Publishers.
Wheelock, F. (1974). *Quintilian as educator*. New York: Twayne Publishers.

Chapter 4
Observing Economics: The Rhetoric of Data, Models, and Statistics

What do we "see" when we observe? What do economists do with "data"? What do economists create when they create models? And what do economists create when they "work with" statistics? That I open this chapter with these questions and the way I phrase these questions implies that the act of observation is an act of education and data, models, and statistics are human creations. I am implying that what economists do is a rhetorical act, not a ploy, but an act of persuasion to convince other economists and non-economists that what economists observe and how they organize "evidence" and construct "truth" are trustworthy and reliable ways of seeing the world and in some cases, everyone should believe, the only way to see reality. Economics at its heart is a rhetorical discipline like history, political philosophy, literature, anthropology, physics, and other fields of knowledge. This does not mean economists are just making up arguments in order to side with a specific ideology, therefore, if we only expose the rhetoric we will expose the ideologues. Ideologues, neoliberals in this case, do exploit economists' work to spew their dogma upon the earth and limit the possibilities of peoples throughout the world, but it is not a given that neoclassical economics is by its nature a neoliberal discipline. I believe that if we understand the rhetoric of economics we can better equip ourselves to challenge the neoliberal dogma that haunts the world and dominates economic thought. I also wish to show that economists not only persuade others of the "truthfulness" of their theories, data, models, and statistics through the use of words, or traditional rhetorical strategies, but also through the use of things, objects, and numbers. Whereas the relationship between neoliberalism and neoclassical economics is a story of affect or how ideologues connect two things (i.e., conservative policies with economic theories such as the Laffer Curve) that are naturally not connected, the relationship between economists and their creations is an act of object oriented ontology. I wish to focus on the act of observation first since I think this lays a good foundation to understand how economists use data, models, and statistics to construct persuasive arguments. Then I want to discuss the rhetoric of data, models, and statistics and how they are often used in economic arguments. I will also show how these concepts are historical constructs and serve

© Springer International Publishing AG, part of Springer Nature 2018   75
J. A. Weaver, *Science, Democracy, and Curriculum Studies*, Critical Studies of Education 8, https://doi.org/10.1007/978-3-319-93840-0_4

powerful disciplinary and political functions. To highlight my points I will rely not only on historians and philosophers of economics but also on specific economic approaches such as Milton Friedman's anti-realist approach, Kenneth Arrow's General Theory of Equilibrium, and a well-known model, the Rational Man Model.

4.1   Observing and Seeing

The Historian of Economics Harro Maas and the Philosopher of Science Mary Morgan phrase the issue of observing this way in their edited work *Observing the Economy: Historical Perspectives* (2013, 1): "What are economic observations? 'My statistical data, of course.'" Their short anecdote demonstrates the way economists are educated to see and think and how evidence is rhetorically and instinctively reduced to data, of course! Learning to observe and see is an act of education. It is not only the disciplining of the eyes and the mind's eye to see a certain way but it is also an external transformation process in which all that is "seen" is immediately translated to mean a specific thing and to exist a specific way. This training in seeing is why some economists like Gary Becker can see the world and proclaim that everything is an economic entity. It also explains why even humans are reduced to resources and human "resource" managers are considered important executives, almost as important as chief executive officers and chief financial officers. To a "well trained" economist everything is data and everything is an economic entity. This "seeing" however marks an important shift for economists and societies governed by economics. As Maas and Morgan (2012, 1) note there is a difference between looking and seeing. Looking like observations deals with "how economists of the past have focused their attention, have found out places, people and practices in which economic behavior might be manifest, and of different perspectives that might be taken to these." Seeing is a "describing, typifying, mapping and maybe even measuring those things that had been observed in appropriate recordings." Looking here implies a chronicling or an acceptance of what is, an anthropology of economics if you will, and seeing connotes a transformation of something that might be economic activity or might not be if the transformation cannot be quantified in a manner that structures the activity and thereby classifying it as an economic activity. This education in seeing explains why Maas and Morgan's hypothetical economist at the beginning of their chapter said data of course. They could have easily extended their retort and said "data, of course, isn't that what everyone else sees and isn't it obvious to you?" To see is an act of quantification, classifying, organizing, and transforming. Once one learns to see, one also learns to observe differently. Observing is no longer an act of proclaiming what is and reporting it or what Maas (2011, 220) says in another work "observations no longer [include] the synthetic activity of reading, note taking, and thinking in one's study." Now observation means "using statistical data sets with the aim of giving precision to mathematical theory." To put it another way Maas and Morgan (2012, 7) note that "quantitative measurement is a stepping-stone in constructing an observation rather than observations leading to quantitative measurements." The model and the "data" shape what one observes and how one observes is shaped by how they are disciplined to see. What this way of seeing has created is an anti-realist approach that simultaneously has witnessed the rise of economics as an important field of knowledge based on projections, closed world models, mathematical equations, and, of course, data sets, but at the same time leading many economists from Tony Lawson (2003), D. Wade Hands (2001), and Uskali Mäki (2002) to proclaim that economics struggles to explain anything in the real world. As Mäki (2002, 3) notes "driven by methodological values that have little or nothing to do with the goal of delivering truthful information about the real world—values such as mathematical elegance and professional status—. They might say that while economics may be the queen of the social sciences in regard to mathematical rigor, it is a failure in so far as its contact with the real world is concerned." There are many ways to approach this dilemma and paradox of economics. How can a field of knowledge be so important to everyday life and policy making yet say almost nothing about real life? How can such a realistic approach to understanding life be so anti-realist? I want to begin addressing these questions by talking about what Maas and Morgan call articulating spaces.

Like most work in science studies today, Maas and Morgan draw their notion of articulation from Latour. As mentioned in chapter two, Latour in his work describes how scientists take something from the "natural" world and turn it into something that can be probed, analyzed, and classified so some kind of proclamation can be made about the meaning of a sentient being or object studied. Articulation then is a process of transformation from one world to another, one language to another, one culture to another. For economists the articulating space is where entities of all kinds are turned into economic objects that can be classified, modified, restructured, and most importantly turned into a datum as part of a data set in order to draw some conclusion that other economists can agree is a statistically and economically significant piece for a theory or policy. How these articulating spaces are constructed and communicated to others is, as Maas and Morgan note, an important matter of trust. In early modern Europe, trust was very much shaped by the political and cultural mores of the time therefore the eyes of an aristocratic gentleman could be trusted to verify experimental facts that could be used then to establish a scientific fact of truth. Times have changed of course. Aristocrats are out, democracy is in, for now. So how does one establish trust in a democracy? The eyes of working people, women, and non-Europeans can now be trusted to see a truth and report it, but eyes are no longer trusted as much as when they were gentlemanly eyes. Trust "became invested in the rigor of mathematical theory, statistical models, and the promises of the computer" (Maas and Morgan 2012, 16). To trust the data one need only know the reliability of the design and the acceptance of a mathematical model and see the numbers generated to know they are trustworthy. To learn how to see and observe has become also an educational matter of knowing what, how, and who to trust.

Instruments, of course, also play a role in the articulating spaces of generating trustworthy data sets and include computers, graphs, and other objects of data generation. These objects can be also referred to as means for transformation in which information of all kinds from historical notations such as birth rates, life expectancies, and suicides, to rainfalls, crop yields per acre, and corn prices can be turned into data. It is the instruments that allow the economists to see "clearly" so something trustworthy can be created and used to articulate sound economic theory impacting government policy.

As with anything there is a price for constructing articulation spaces as economists have since the rise of neoclassical economics in the late nineteenth century. It has narrowed the field of economics at the very time some economists are making bold proclamations of covering everything. It has reduced life in the name of limited vibrancy and reliability of knowledge, and narrowed notions of research. Benoît Godin in his study of measurement and statistics especially in connection to the Organization for Economic Co-operation and Development (OECD) suggests that there has been a hardening of the research imagination. Godin (2005, 58) defines the term research use to mean "an 'act of searching closely and carefully,' or 'intensive searching.'" Now it mostly is referred to what disciplined scientists do and usually only for those scientists who approach their research with an application in mind, but as Godin notes the traditional meaning of research could apply to anyone who searches closely, carefully (systematically), and intensively at any topic of interest. In economics, it was not uncommon for economists to use countless forms of information for their research including historical documents, travel reports, observational notes, government reports, and novels (Maas 2011; Maas and Morgan 2012, 3). Now, economic evidence is a matter of data. Maas (2011, 218) marks this distinct change as a major shift "from one which the political economist weighs different sources of evidence, to one in which statistical data collected…came to count as the one and only homogenized source of observation." With the "maturing" of economics we see the narrowing of the field. It marks a rise of standardization not only of the field of economics but also of the world in many ways. This standardization hides, often, more than it reveals. As Martha Lampland and Susan Star (2009, 21) write standardization hides "embedded biases in representations of knowledge, both blatant…and subtle (e.g., in the categories in databases)." They also suggest that standardization "presumes the ability to constrain a phenomenon within a particular set of dimensions, as well as the ability to dictate behavior to achieve the narrowly defined dimensions that stipulate its outcome (Lampland and Star 2009, 14). Later we will see this is exactly what happened when the OECD started to standardize their statistical approaches and limit notions of science and technology to economic constructs. We can also see the same phenomenon happening with the use of the term "data" as well. Daniel Rosenberg (2013) in reports that etymologically the word data means that which is used for argumentative purposes, something that is used for foundational purposes to make a rhetorical point. Data is at its origins rhetorical, but in the twentieth century data's rhetorical function and history has been embedded and camouflaged into a more precise and limited meaning that defines data as "information in numerical form" (Rosenberg 2013, 33). Of courses, everyone knows data is not rhetorical! So goes the rhetoric we are told.

4.2   The Rhetoric of Data

How did societies get from the Latin origins of the word data Rosenberg mentions to the current status of data as more than a foundation of given assumptions to construct an argument? Rosenberg's (2013, 18) etymological trip reveals that the term appears in English in the seventeenth century and its Latin root means "something given in an argument, something taken for granted." This meaning implies two things that are important to note. The first is that in presenting an argument in order to persuade someone means I can use any data I deem necessary to support a claim but it does not mean I by any means speak the truth. When I present something to support an argument it does not mean I am speaking truthfully, just presenting something to support my cause. Yet the second point to note is indeed when I am making a case and assuming to take something for granted I am making something like a truth claim. At the very least I am offering up a hypothetical and saying something to the effect "let's assume…" or "would you not agree with me that…" As soon as we assume or you agree with me that… I have you. I have you in my world and in my world my data rules by definition. Data, Rosenberg (2013, 18) then suggests, is different from facts and evidence since "facts are ontological, evidence is epistemological, data is rhetorical." The difference between a fact, evidence, or truth is "when a fact is proven false, it ceases to be a fact. False data is data nonetheless" (Rosenberg 2013, 18). Today, fact, truth, evidence, and data have become more intertwined. When an economist responds to the question of what are economic observations with "my statistical data, of course," the rhetoric of data remains, mostly hidden or erased, because it is assumed that the data create the evidence, reveal the facts, and establish the truth. From a rhetorical stand point the data only establish the parameters for debate, from an economists perspective data establishes everything. As Rosenberg (2013, 22, 28) notes about the current state of data we are "swimming" in it and "the future is data." Yet, there is still one important piece that is missing. Although facts, evidence, and truth have become more intertwined with data it does not mean the intermingling is comfortable or neat. In spite of its closer connection to truth claims, Rosenberg (2013, 37) notes "Data has no truth. Even today, when we speak of data, we make no assumptions at all about veracity….It may be that the data we collect and transmit has no relation to truth or reality whatsoever beyond the reality that data helps us to construct." Geoffrey Bowker (2013, 170–171) agrees with Rosenberg on many fronts when he writes "If you are not data, you don't exist" and "We are managing the planet and each other using data [hopefully in a factual and truthful manner]…What we need is a strongly humanistic approach to analyzing the forms that data take…which enables us to envision new possible futures." While everything is becoming data and anyone can now make a truth claim say against vaccinations or evolution because data is assumed to be the same as fact, evidence, and truth, Rosenberg and Bowker offer a different story. We may be swimming in an endless ocean of data filled with threatening (loan?) sharks and other predators and our identities may be constantly constituted by data, it is still not the same thing as fact, evidence, truth, or reality.

These things have to be constructed. The environments, the structures, the logics, and the foundations for thinking, acting, and believing have to be invented. There is great work to be done. The data has to be "scrubbed" and manipulated to fit a certain way of thinking, world view, or computer generated model of economic behavior. It is here where economists live and it is here where the rhetoric of economics now resides; in the art of scrubbing. In their work on the early United States neoclassical economist Irving Fisher, Kevin Brine and Mary Poovey demonstrate this important point. Fisher, like other early neoclassical economists, was interested in creating a theory of utility or the idea "that, under idealized [always idealized which is to mean never realistic] conditions, the prices that economic markets establish reflect an exchange equilibrium, or balance between the amount of buyers' desire for goods or services and prices sellers want for those goods or services" (Brine and Poovey 2013, 62). In order to make his case, Fisher did not just compile some "data" run a test or place it in a neat looking graph, he married "theory to empirical data in a quantitative form, Fisher created a framework that could fit the theory to the data" (Brine and Poovey 2013, 72). He scrubbed the data until it was able to help him construct a reality conducive to his ideas regarding utility.

Our role in the scrubbing of data is not to "dry" the data off as if we have no other role to play in the construction of reality. Instead, in the name of our very meanings and existences, we are to contest any data driven construction of who we are and what our future may hold. That is to say, when an economist emerges from his economic observations of a data-constructed reality and proclaims that human beings are "rational men," "Optimizing actors" or "selfish pleasure seeking economic agents seeking the right price for an item and living an idealized state of equilibrium," our role as something more than rationalizing individuals whose emotional range and motives for living moves far beyond these simple and limited choices is to challenge these assertions; these reality construction data points and ontological and epistemological foundational assumptions. The truth claims of economists are no more given than any other "new possible futures" Bowker mentions. This does not mean, at least for me, that we should embrace the latest conspiracy theory, the anti-vaccination movement, or the creationist fundamentalist who think Jesus lived shortly after dinosaurs. I would hope we have better and more interesting worlds to create. In order to challenge the limits of economists' data constructions we need to enter their worlds, understand their rationale, and offer alternative ways of knowing and being. This means the data constructed by economists needs to not only be critiqued but coopted. It also means non-economists have to construct their own data sets in order to offer up alternative pathways into the future. These data sets can be mathematical and quantifiable but need not be limited in this way. If data is everywhere and we are swimming in it then the data constituting life consists of much more than mathematical and quantified forms. Economists might balk and the powerful may laugh and ignore the alternative worlds created from more than mathematical and quantified data sets but their reaction is a power play build on a foundation of insecurity and Nietzschean Ressentiment. With some imagination, economists' monopoly on knowledge production and governance can be curtailed where at the very least they play a lesser role in constructing reality and naming truth.

This story of the ubiquitousness of data is not finished. There is still numeracy or quantification, power, ideology, and Derrida's Fever to cover in this section. The Historian of statistics Theodore Porter (1995) in his book title *Trust in Numbers* alone tells a story of why numbers or quantification are important. People today in many parts of the world trust numbers over "mere" stories people tell, as if numbers do not tell stories. In his novel *The Life of Pi*, Yann Martel tells the story of a young boy Piscine Patel or Pi for short since Piscine sounds too much like pissing, and his harrying experience of surviving a sinking ship on a lifeboat with a Bengal Tiger for months. When they finally land on the shores of Mexico the Tiger, aptly named Richard Parker, walks away somewhere never to be seen again and Pi ends up recovering in a hospital where he is visited by two insurance officials who wish to settle the case if money needs to exchange hands. Pi relays the story to these officials who think he is delirious from the experience, maybe even mad. How could a boy survive on a lifeboat with a tiger? Impossible. Pi sees that his story is not acceptable to the insurance minded so he tells them an alternative story that they can believe. His first story was an allegory. The non-survivors from the lifeboat, his mother, was an orangutan, a Taiwanese sailor a zebra, the cook a hyena who killed the other two. And the tiger? Pi was the tiger, he killed the hyena. Now there was a story they could believe because as Pi pointed out earlier in his alternative fictional account "I know what you want. You want a story that won't surprise you. That will confirm what you already know" (Martel 2002, 336). The insurance claims adjustors needed something they could trust. They needed a story that could confirm what they knew: a ship had sunk, there was only one survivor, and this story needed to be reduced to numbers so it could be decided if a claim was justified. If the ship sank by accident then what number was due the owner, if it was sabotage or neglect then was there a number greater than zero that was due? The life of adjustors is always reducing meaning to a number. Yet, what the fictitious adjustors did not know or had completely forgot because Pi's story did not fit the ways they were inculcated in the ways of numeracy, they needed to read Porter and be reminded that "Numbers, too, create new things and transform the meanings of old ones" (Porter 1995, 17). When numbers represent data new ways of seeing and thinking are invented and economists lead the way along with natural scientists, engineers, and insurance adjustors who invented actuary tables to predict the future, maximize profits and limit payments.

Numbers came to dominate these fields of knowledge because they offered an illusion of stability and certainty. It gave scientists, economists, engineers, adjustors, business people and political leaders a peace of mind or a rhetorical escape hatch from any debate that challenged their judgments as sound and rational. Numbers, especially in the form of data, allowed these purveyors of trust to present "evidence" to the soundness of their decisions and the truthfulness of their policies. I join Elizabeth St. Pierre (2013, 223) when she states "I'm interested in the occasion of data's appearance; that is, in when, where, why, how, and by whom data I called into being to do some work." When economics quantified and mathematics dominated the field after World War Two what was data called to do? What work was it asked to do? The answer lies initially in only a handful of interested people in the budding field of economics such as William Jevons, Richard Jones, Albert Marshall, Irving Fisher, Vilfredo Pareto, and Francis Edgeworth. They were the early pioneers of quantifying economics. Before their push to turn political economy into a mathematical science, the general consensus was represented by John Stuart Mill who firmly believed that the study of economic interactions were too complex to be able to reduce them to mathematical formula without seriously eliminating important aspects of economics. This is why Mill did not consider economics a science like physics or astronomy. We already see that Fisher countered Mill's objections by creating data that could show how utility or equilibrium functioned in economic theory and Jevons created what Maas (2005) refers to as the graphical method. This method for Jevons would challenge Mill's dominance in economics by eliminating the divide "between political economy and statistics" (Maas 2005, 233). As Maas (2005, 236) points out Jevons use of graphs "was not a search for the best fit of the graph; it was a first step to the best explanation." Data on the graphs were the starting points to uncover phenomenon. "Phenomenon," for Jevons (Maas 2005, 234–235), "could then be further analyzed to reveal the natural laws which they obeyed. For Jevons, these laws were stable functional relationships....it is evident that his general approach to the sciences did not involve a split between pure theory and statistics, but rather was motivated by a unified framework in which analogical reasoning played a dominant role." Data then were called upon to serve statistics which could be used to discover the universal laws of economics thereby making economics similar to physics and astronomy. Data's appearance marks the beginning of transforming economics from a philosophical and moral discipline to a science. Numbers marked the beginning of stabilizing economics just as they stabilized physics and astronomy to make more general and sweeping claims about the scope and relevance of its field.

Alain Desrosieres (2009, 312) raises a similar question to St. Pierre's when he asks "what is the purpose of the very act of quantification and whom does it serve?" Like St. Pierre, Desrosieres' question raises issues of power. To what power structure do data owe its allegiance? Of course, if data is meaningless as Rosenberg and Bowker suggest then data owes its loyalties to no one and nothing, but this cannot explain the obvious connections between data and power. From the beginning of the development of statistics and numerical data, state powers understood the importance of these tools for meaning construction. The word statistics itself refers to state officials who collected information in order to control and govern the populace. Desrosieres (2009, 318) points out that "political unification and statistical unification go hand in hand." From the beginning then, statistics and numerical data were connected to governing powers. "Quantification serves to provide tools for comparisons," Desrosieres (2009, 312) suggests, "to coordinate them by standardizing them, and to control and stimulate players by ranking their performances on normative scales." What areas of the kingdom are more likely to be filled with crime, where should the police force be located, which units in the military are more effective and combat ready, where are the suicide rates higher and why, what government officials are more efficient in carrying out their duties, which schools are better performing schools to get students ready for the job market? All of these are questions raised by institutions of power in which each question can be quantified to give a streamlined account of who and what performs best. Data serve state powers to evaluate, judge, condemn, deploy, honor, disgrace, monitor, regulate, and promote its population. Data remind us all that we live in a society as points on graphs, tables, reports, and numerical markings of all kinds. When economists became more and more connected to governments it should be no surprise it happened after they developed more detailed statistical means to turn economics into a science. Once statistics became more essential to economics and data were constructed, states began to see the value of economists. This codependency between economists and governments does not come without friction. In the last section when I discuss statistics in more detail I will cover the tensions that exist between statistics and governments.

4.3 Data as Archive

It is hard to contest Rosenberg's claim that the future is in the data and we are swimming in it or Bowker's that if you are not data you do not exist. There is a strong truth to their assertions. At the same time, I can, as Derrida has, argue the opposite. That as long as we are swimming in data there is no future and as long as we are data there is no existence. Derrida (1996, 2) tells us that "the meaning of 'archive,' its only meaning, comes to it from the Greek arkheion: initially a house, a domicile, an address, the residence of superior magistrates, the archons, those who commanded... On account of their publicly recognized authority, it is at their home, in that place which is their house (private house...), that official documents are filed. The archons are first of all the documents' guardians....They have the power to interpret the archives." With Derrida's lesson on the Greek origins of the term archive we see the early connections between government and statistics. The holder of these documents has the power to confer meaning over what the documents mean; to declare who exists and who is a non-entity. "The archontic principle," Derrida (1996, 3) comments, "of the archive is also a principle of consignation, that is, of gathering together." Isn't this what a statistician does? He or she compiles data for government, economic, and educational institutions so decisions can be made, policies created, degrees conferred, and lives constructed? Then, Derrida (1996, 4) declares "there can be no political power without control of the archive, if not of memory. Effective democratization can always be measured by this essential criterion: the participation in and the access to the archive, its constitution, and its interpretation." Is this why ever great power has a library? Ancient Rome Alexandria, Medieval Europe monasteries, Napoleonic France new universities, and the United States the Library of Congress? Did they form these archival bodies in order to control the memory and determine participation and access parameters to this memory?

This drive to control memory, participation, and access is both a matter of the past—what one said, what one did, who was—and a matter of the future—how one will be remembered, if at all (hence my dedication to Jimmy Geiger), and how one will be reconstructed by others, if at all. The archive simultaneously strives to keep the past alive and assure the possibility of the future. Yet at the same time, Derrida points out, the archive seeks to eliminate both the past and the future. The goal of the archive is to preserve something from the past, but as soon as that something is preserved the past is eliminated and reduced to that preserved unit. The archive destroys the past by its very act of preserving or housing the past. What remains is what is remembered and how educational and political institutions are structured determines how what remains is remembered in the future. Derrida notes that the act of archiving is founded on an act of violence. That which is preserved and housed safely for a future unknown destroys the past and in every case eliminates most people and non-human sentient beings from existence. These beings died and were buried but they were erased when the archive preserved only a piece of them or nothing at all. As Derrida notes in Naas's (2015, 136) book *The End of the World and Other Teachable Moments: Jacques Derrida's Final Seminar*, "The archive drive…is an irresistible movement not only to keep traces but to master them, interpret them." The drive to preserve and master controls the future by eliminating the past. This I want to suggest to you is exactly what data does. Data surely marks who we are to ourselves, to government agencies, educational institutions, and economic entities, but it also eliminates who we are. Data transforms us from who we are to archived documents in the form of what grades we earn, what income we make, what products we buy, and on and on the data consumes us. There is a drive for governments to collect data because there is a drive to master individuals, to dictate who shall live and who shall die, who shall have a future and who will not even have a past. The same is true for economic entities who seek to master individuals, to determine who is a human being and who is a worker or consumer, who is alive and who merely are points on economic forecasts. Data are archives that violently construct who people and non-human entities are, and what objects are. This is the core question surrounding the rhetoric of data and why economic data is important for curriculum scholars to pay attention to. Our future and past existence depends on it.

4.4   Modeling Rhetoric

In her work *The Dappled World*, Nancy Cartwright (1999, 1) writes what to many of us is the obvious, "we live in a dappled world, a world rich in different things, with different natures, behaving in different ways." Yet she has to state the obvious because her focus is on physics and economics. "The dappled world," Cartwright continues, "is what, for the most part, comes naturally: regimented behavior results from good engineering." In other words, the world is messy perhaps governed by an order but the "laws that describe the world are a patchwork, not a pyramid. They do not take after the simple, elegant and abstract structure of a system of axioms and theorems." The models physicists and economists invent create this order. It is an order imposed on the world. It is imposed by "disciplines with imperialist tendencies" (Cartwright 1999, 1). What she means by imperialist tendencies has nothing to do with political desires. Instead she is referring to the belief of many physicists and economists that their models can explain everything. And indeed their models can explain everything! With one exception, of course, the world must fit into the nomological machine called a model. Anything that does not fit is eliminated from reality. The nomological machine, as Cartwright (1999, 50) calls a model, constitutes the creation of a system that "is a fixed (enough) arrangement of components, or factors, with stable (enough) capacities, that in the right sort of stable (enough) environment will, with repeated operation, give rise to the kind of regular behavior that we represent in our scientific laws." The problem is not with the nomological machinery it is the exuberance to transcend the limits of the model in order to make universal, unsubstantiated claims about the reach of the model to explain reality. More importantly, I think Cartwright (1999, 17) taps into a deep "yearning for 'the system'" and a "faith that our world must be rational, well ordered through and through." Models, then, serve not only a scientific purpose of establishing empirical facts or confirming an observation in reality, they serve a rhetorical function to construct reality—a reality that exists only under the right conditions. Models also serve intellectually political purposes by establishing which voices and perspectives policy makers should listen to or what models ideologues should use to claim a certain economic theory is incontrovertible. This is what neoliberals have done with economics. They have taken a few theories like the Laffer curve proclaimed its universality and constructed a political system and public mindset that there are no alternatives and no way around the economic laws of capitalism. Models are constructs that in turn become powerful constructors of reality even though the models rarely if ever fit into any kind of reality other than the one that is assumed in the models. Models have become tools to establish intellectual authority, direct policy, and limit ways of seeing the world. They have become the main way economists work. As Mary Morgan (2012, 2) writes economics "is now a very different kind of activity....By the late twentieth century, economics had become heavily dependent on a set of reasoning tools that economists now call models...that can be manipulated in various different ways." So why have models become so prominent in the field of economics and other fields such as astronomy, physics, and biology? An obvious reason is they are very good at producing evidence and knowledge. There are, however, many other reasons.

The philosopher of biology, Evelyn Fox Keller (2002, 158) asks "What is a model (or theory) a model (or theory) of and what is it for?" The immediate response is it is for generating hypothesis or data (under the right conditions) that can be used to understand a phenomenon in reality. Models are also used to predict outcomes. Of course, in economics in boom times the predictions are always praised and in times of depression or recession the economists are asked why did you not see this coming? In matters of prediction, models are awful indicators of what is to come when economists are asked to explain a downturn. Economists still have not provided adequate explanations as to why "the economy" collapsed in 2008. It is probably because models do not take into consideration predatory actions, corruption, insider information (most models assume a neoliberal notion put forth by Hayek that information flows incompletely but freely in a free market system), graft, and other realities of life on Wall Street and in other economic centers in the world are erased from models. There are other reasons that are just as important and meaningful in creating a model. Models spark the imagination. As Keller (2002, 158) notes metaphors are "a class of models (analog models), both serving the same heuristic value of redescription." The most famous metaphor in economics of course is the "invisible hand" that was generated to explain the freedom of decision making in a free market and also to explain away any problems capitalism could not address. There have been others as well including Reagan's metaphor of a rising tide raises all boats to simultaneously support the notion that redistributing wealth from the middle class to the rich will spread wealth to everyone and to deflect any criticism that the theory creates income inequality. Perhaps most importantly, models are generators of fiction. A model inherently, no matter what assumptions are used to generate the outcomes, creates possible, real worlds that are nonetheless always fictive. Morgan (2012, 280) compares two fictive worlds when she says "the laboratory scientist creates a controlled real world within an artificial environment while the modeler creates an artificial world in a model." Like a novel, economic models can tell us a wonderful story, but it is still, we must always remember, a fiction just as the controlled real world of the laboratory scientist is. Keller reveals one reason why models are so pervasive at least in her own field of biology, but I am sure it fits other fields of knowledge as well. "Need," Keller (2002, 202) "is undoubtedly the major impetus for the increasingly favorable reception of mathematical and computational modeling. But presentation is another factor. New experimental results can now be represented in a format that is accessible and persuasive to an audience." In other words, models serve rhetorical purposes.

Traditionally there are two types of models used in economics. The first model is created from mathematical formalism and is used to predict what will happen and the second is also based on mathematical formalism but is used to create hypotheses or theories. The first is interested in connecting the model to the so-called real world and the second is interested in whether the model fits the logical parameters established by the mathematical principles. It is this second model that Friedman made famous in his 1953 essay "The Methodology of Positive Economics" in which a model is deemed significant if it is consistent with its own logical principles. The outcome of these models then for economic policy is not important because they reveal something about the reality of an economic phenomenon but because they are consistent with the logic undergirding the model. Often labeled as anti-realist, the Friedman approach to economics has impacted economic and political policy greatly simply because the outcomes are deemed logically consistent. This approach to modeling and policy making serves as the best example of Cartwright's point about the limits of models and the imperialist tendencies of economics. Just because something is logically consistent does not mean it has any application or connection to any part of reality. The best example of the problems this second approach to modeling creates is the general theory of equilibrium, health care, and scientific knowledge. The general theory of equilibrium is, under perfect conditions, "the state of the economic system at any point of time as the solution of a system of simultaneous equations representing the demand for goods by consumers, the supply of goods by producers, and the equilibrium condition that supply equal demand on every market" (Arrow and Debreu 1954, 265). When neoclassical economics and neoliberal ideologues use the term perfect conditions they are usually referring to four things: There should be no government interference in the market (except for neoclassical economists like Arrow when the market fails to perform), no corporate monopolies, no worker organization or corporate collusion, and free flow of information no matter how incomplete. Neoliberals in the last 30 years have stood firm on insisting on no government involvement and no trade unions but have turned a blind eye to corporate monopolies. Neoliberals and some neoclassical economists have tried to apply this theory to every aspect of life when even one of the creators of the theory, Arrow, insists that there are at least two areas where the general theory does not apply, health care and scientific knowledge. Arrow (1962, 1963) in 1963 for health care and in 1962 for scientific knowledge demonstrated that the theory does not apply. In other words, perfect market conditions do not occur in these two fields and some form of intervention from regulating bodies, government agencies, union representation is warranted and necessary. Neoliberals have ignored the limits of the general theory of equilibrium model and forced health care and scientific knowledge into these strict parameters working from the assumption that all economic systems whether they deliver health care, produce knowledge, bake bread, forge steel, or manage data are all the same and fall under the dictates of the general theory of equilibrium. This entrenched and blind ideological fervor has created great problems for health care in the United States and greatly damaged public schools and universities in the United States as well. This neoliberal fundamentalism stems from the fictive world created by the general theory of equilibrium and rhetorical fervor that perfect conditions as defined by neoclassical models are accurate measures of what can be found in reality or even worse can only be found in the models but that is good enough justification to persuade others to demonize government regulations as inherently harmful, to embrace corporate mouthpieces like the chamber of commerce, reject unions, and pretend information flows freely. This model of perfect competitive conditions has shaped and continues to shape the collective imagination of people throughout the world and it has promoted flawed policies and created hostile conditions for workers. As Mary Morgan (2012, 218) points out, as with all models the outcomes from the application of a general theory of equilibrium, not as created by Arrow and Debreu but espoused by neoliberals, is based on the story that neoliberals tell themselves, politicians, policy makers, and the general public and "stories can shape the reasoning resources of models before going on to show how and why economists working with models typically ask questions and tell similar kinds of stories when they reason with them." Neoliberalism with the help of manipulated economic models has become a self-fulfilling prophecy with disastrous results and consequences.

Another famous economic model, the "rational man" model, demonstrates how models are reductively used to support neoclassical assumptions. Originally referred to as "economic man," "Homo economicus," Jevon's "Calculating man" or eventually using its current name Prisoner's dilemma, the rational man model is as Morgan (2012, 137) points out "not a full man." I would say it is not a full human being. It is emotionally stunted since its primary emotion is selfishness or self-preservation. In classical terms "the motivations" for "the rational man" "consist of one constant positive motivation, namely, a desire for wealth and accompanied by only two 'perpetual' negatives: the dislike of work and the love of luxuries." Other than explaining the laziness of rich people this stunted version of a human being does not explain the reality of working conditions or those like Donald Hall (2003) who live to work no matter how demanding it may be. In Frank Knight's model it is assumed that the "calculating man" has "full information about everything in the economy…and perfect foresight about the future" (Morgan 2012, 150) so he can make the right choices about what decisions to make and when. As Morgan (2012, 152) notes Knight constructs human beings as "purely impersonal utility maximizing agent[s]…a pleasure machine." The "rational man" is one dimensional and this one dimensionality creates a dilemma, a prisoner's dilemma.

Prisoner's dilemma is a version of the "rational man" model. It creates the scenario that there are two suspects of a crime in two separate rooms. They are both presented with choices and each without knowing what the other is doing or thinking. As the scenario goes the district attorney "points out to each prisoner that each has two alternatives: to confess to the crime the police are sure they have done, or not to confess. If they both do not confess, then the district attorney states he will book them on some very minor trumped up charge such as petty larceny…and they both will receive minor punishment; if they both confess they will be prosecuted, but he will recommend less than the most severe sentence; but if one confesses and the other does not, then the confessor will receive lenient treatment for turning state's evidence whereas the latter will get 'the book' slapped at him" (Morgan 2012, 350). The beauty of this dilemma is no matter what the prisoners choose it proves that neoclassical thinking is correct: all individual decision making is rationally geared towards self-preservation. This becomes a prisoner's dilemma for anyone who wishes to challenge neoclassical economics and neoliberal ideologues since no matter what one chooses one ends up supporting an argument for neoclassical economics. I think this is one reason why critics of the limits of economic modeling like Mary Morgan and Nancy Cartwright critique the limits of modeling and the logic undergirding it. It is a way around the bounds of the model and by standing outside the model they are able to challenge its presuppositions.

Another way to challenge models such as the "rational man" model is not to focus on the rational dimension but on the man. The history of economics without a doubt is a gendered subject. For all of the nineteenth and most of the twentieth century women were not considered to be economic agents and most economists have been and still are men. This means the models used to study economics will reflect and embody the values and beliefs of men. Historically this has meant that what was often referred to as "women's work" was not valued as a topic to study and the primary focus on "men's work" dominate the models developed to understand what economics meant and studied. As a result, we can conclude that in the fictive world of modeling the actors are always rational, bent on self-preservation, avoiding as much work as possible in their drive for pleasure, and most definitely male making the rational man model one of the most influential examples of how the rhetorical figure of the universal western man still functions to define what it means to be human.

4.5   Statistical Rhetoric

In his closing remarks in the *British Society for the History of Science* issue on transnational and science studies, Pestre (2012, 427) makes this seemingly banal statement: "Through statistics, scientific management, social engineering and operational research, modern science also quickly became a way to govern. Historically, it has always been a way of helping those in power to have a better understanding of the world, and thus to better manage both nature and society." Pestre's statement is tame in that his word choices suggest nothing about manipulation of statistics or knowledge to force nature and society into a certain view that inherently justifies and sanctions the rule of certain people over others, or in today's case multinational corporations over people. His words suggest that governing officials and bodies use all rational and intellectual means to make important decisions for the betterment of nature and society. Godin (2005, 296) is succinct and to the point "governments produced statistics in order to control populations." And Desrosières (2009, 318) makes this observation "political unification and statistical unification go hand and hand. This tie is implicit in the very etymology of the word statistics: the science of the state." In order to convince governing officials and eventually "the public" of the reliability of statistics a certain rhetoric had to be created to justify the validity of statistics to help government decision making and along the way help unify people in order to control them so the entity called a nation and today called a multinational corporation will be sanctioned as the legitimate mechanism for governing. As Gigerenzer et al. (1989, 122) note in their book *The Empire of Chance: How Probability Changed Science and Everyday Life*, in the Ancient Regimes of early modern Europe statistics were accumulated but there was no need to assure the reliability or objectivity of the data collected and compiled. "But experts in modern democracies…must be armed with 'objective' inferential tools and mechanized experimental set-ups. Numbers and methods of manipulating them are crucial to their authority." With the rise of democracies began the process of developing a rhetoric to justify not only the use of statistics to make decisions but to assure "the public" that the statistics pointed the policy makers in the direction they wish to follow and the data are reliable because they are objective and neutral. The history of statistics then is not only about the development of modern statistical methods that continue to be used today in some form, but it is also a matter of how statistics developed a rhetoric of acceptability and reliability to inform policy decisions. It is then presumed to be a coincidence that the meaning of statistics and the will of the rulers coincide. Statistics have become a firewall between decisions of the powerful and criticisms of those decisions.

If Desrosières' mantra is correct that political unification goes hand and hand with statistical unification, then it is also correct to say that the rise of statistical thinking and decision making went hand and hand with state support of academic research. Godin (2005, 263) notes that "the concept of basic research acquired political stability (partly) because of statistics. The latter helped academics and bureaucrats to convince politicians to fund research." The history of basic science research with the rise of statistical thinking and the modern democratic state, marks a major shift as to where it was done. Prior to the democratic era of governance basic research was done in some universities but most was done in two other areas: aristocratic organizations and royal courts. Science as it developed in the early modern period was conducted by wealthy independent gentlemen such as Boyle who could fund their own research and create their own organization to verify their findings such as the Royal Academy of Science. If one were not of the right family bloodline or independently wealthy, a natural philosopher could seek out the support of a royal family as Galileo did numerous time. Each approach to supporting basic research had its own politics and rhetoric. In order to turn his findings into legitimate scientific achievements Boyle had to seek out and earn the approval of his fellow aristocratic natural philosophers and Galileo had to consistently perform in front of the Medici family with new found discoveries named in their honor like the moons of Jupiter in order to earn continued financial support. In this sense the university currying favor with political entities is not very different from the early modern period of science. The difference is where Boyle's notion of truth was wrapped up in the eyes of aristocrats to label something a fact and Galileo needed monarchical support, universities needed statistics to seek government funding and approval. As Godin (2005, 285) notes "statistics were influential in helping give basic research political identity and value." Just as it worked for Boyle and Galileo, it worked for universities as well. Not only did states fund basic research in newly created scientific fields such as chemistry and biology but also by the early twentieth century sociology, psychology, anthropology, and economics were well supported for their basic research. This was accomplished at the price of assuring political entities that the research was reliable, objective, and value neutral which is to say statistically reliable.

Since the decline of the nation state and the rise of the multinational corporation in the early 1980s, universities have seen its support from state funding agencies decline and as a result too the value of science and the veracity of scientific claims as well. This decline in state funded support of basic research can explain in part why, in the United States, more people are doubting the findings of science when it comes to such issues as evolution, climate change, and vaccinations. This decline also has required universities to shift their focus and instead of courting state funding agencies, multinational corporations have now become a target for funding basic research. Statistics are still involved in assuring multinational corporations that the research is reliable, but what has changed dramatically is the areas where research is now funded. Instead of the funding of the humanities and traditional science fields such as physics and chemistry, now the focus is industrial agendas such as polymers, pharmaceuticals, and genetics. This has caused a major shift in the meaning and purpose of the university. Now the purpose of the university is not to create citizens of a nation but workers ready to hit the human resources offices looking for a job and fitting the needs of corporations. Students have become customers and universities have become job placement services while researchers have become adjuncts to the needs of multinational corporations while non-essential faculty have become adjuncts of the neoliberal university eking out a living at this institute of job training formally known as an institution of higher learning and at that one. In the end statistics served to legitimate basic research within the university but it could not save it.

The reason why statistics could not save the university from a neoliberal ideology that views everyone and everything as an economic entity is ultimately statistics serve the state, and now the multinational corporation. Godin (2005, 317) in his study of the Organization for Economic Co-operation and Development, an organization whose task was to create reliable statistical information on science and technology to inform policy for any member nation in regards to economic development and education, asked rhetorically "what end, then did statistics serve?" What Godin found is that statistics were used to control public discourse, rationalize policy decisions, and curb dissent. Statistics "helped the decision maker and the politician construct discourses aimed at convincing the citizens about a course of action already chosen or taken" (Godin 2005, 317). For instance, Godin points out that what the statistical work at the OECD did was to confirm the widely held political view that science is an economic endeavor. "In the mid-1970s, scientific and technological policies were increasingly focusing on innovation rather than support for science per se. Consequently, the current agenda of OECD member countries is mainly economic, with technological innovation believed to be a key factor; if not the main one, contributing to economic growth" (Godin 2005, 88). This goes a long way to explain why universities now sell themselves as institutions of economic growth and economists are highly valued as the keepers of the data. Statistics then, in spite of its courting of key ideals as objectivity, value neutrality, and quantification of data, is a deeply ideological endeavor. Again Godin (2005, 6) is helpful and insightful: "statistics are a rhetorical resource used to convince for a course of action already selected. Statistics crystallize choices and concepts that support them." The main issue is not how can we continue to expose statistics as an ideological culprit in governing struggles, but how can statistics be used to open up policy decisions before those decisions are already made and how can statistics be used to move away from the myopia of neoliberal thought that reduces everyone and everything in life to an economic transaction and towards a more democratic mode of decision making and a more artful way of living? This question should be one of the major challenges curriculum scholars address immediately.

References

Arrow, K. (1962/2002). Economic welfare and the allocation of resources for invention. In P. Mirowski, & E.-M. Sent (Eds.), *Science bought and sold* (pp. 165–180). Chicago: University of Chicago Press.

Arrow, K. (1963). Uncertainty and the welfare economics of medical care. *The American Economic Review, 53*(5), 141–149.

Arrow, K., & Debreu, G. (1954). Existence of an equilibrium for a competitive economy. *Econometrica, 22*(3), 265–290.

Bowker, G. (2013). Data flakes: An afterward to 'Raw Data' is an oxymoron. In L. Gitelman (Ed.), *"Raw Data" is an oxymoron* (pp. 165–171). Cambridge, MA: MIT Press.

Brine, K., & Poovey, M. (2013). From measuring desire to quantifying expectations: A late nineteenth-century effort to marry economic theory and design. In L. Gitelman (Ed.), *"Raw Data" is an oxymoron*. Cambridge, MA: MIT Press.

Cartwright, N. (1999). *The dappled world: A study of the boundaries of science*. Cambridge: Cambridge University Press.

Derrida, J. (1996). *Archive fever: A Freudian impression*. Chicago: The University of Chicago.

Desrosières, A. (2009). How to be real and conventional: A discussion of the quality criteria of official statistics. *Minerva, 47*, 307–322.

Fox Keller, E. (2002). *Making sense of life: Explaining biological development with models, metaphors, and machines*. Cambridge, MA: Harvard University Press.

Gigerenzer, G., Swijtink, Z., Porter, T., Daston, L., Beatty, J., & Krüger, L. (1989). *The empire of chance*. Cambridge: Cambridge University Press.

Godin, B. (2005). *Measurement and statistics on science and technology: 1920 to the present*. New York: Routledge Press.

Hall, D. (2003). *Life work*. Boston: Beacon Press.

Hands, D. W. (2001). *Reflection without rules: Economic methodology and contemporary science theory*. Cambridge: Cambridge University Press.

Lampland, M., & Star, S. L. (Eds.). (2009). *Standards and their stories: How quantifying, classifying, and formalizing practices shape everyday life*. Ithaca: Cornell University Press.

Lawson, T. (2003). *Reorienting economics*. New York: Routledge.

Maas, H. (2005). *William Stanley Jevons and the making of modern economics*. Cambridge: Cambridge University Press.

Maas, H. (2011). Sorting things out: The economist as an armchair observer. In L. Daston & E. Lunbeck (Eds.), *Histories of scientific observation* (pp. 206–229). Chicago: University of Chicago Press.

Maas, H. and Morgan, M. (Eds.). (2012). Observations and Observing in Economics. *History of political economy* (Vol. 44, Annual Supplement, pp. 1–24). Durham: Duke University Press.

Maas, H., & Morgan, M. S. (2013). Observation and Observing in Economics. *History of Political Economy, 44*(Supplement 1), 1–24.

Mäki, U. (Ed.). (2002). *Fact and fiction in economics: Models, realism, and social construction*. Cambridge: Cambridge University Press.

Martel, Y. (2002). *The Life of Pi*. Toronto: Vintage Canada.

Morgan, M. (2012). *The world in the model: How economists work and think*. Cambridge: Cambridge University Press.

Naas, M. (2015). *The end of the world and other teachable moments: Jacques Derrida's final seminar*. New York: Fordham University Press.

Pestre, D. (2012). Concluding remarks. Debates in transnational and science studies: A defense and illustration of the virtues of intellectual tolerance. *Br Soc Hist Sci, 45*(3), 425–442.

Porter, T. (1995). *Trust in numbers: The pursuit of objectivity in science and public life*. Princeton: Princeton University Press.

Rosenberg, D. (2013). Data before the fact. In L. Gitelman (Ed.), *"Raw Data" is an oxymoron* (pp. 15–40). Cambridge, MA: MIT Press.

St. Pierre, E. (2013). The appearance of data. *Cult Stud Crit Methodol, 13*(4), 223–227.

When I was hired as a professor of educational foundations at Louisiana State University-Shreveport I was twenty-nine years old and thought I knew something about social justice and a living democracy. I had a lot to learn. I arrived on campus and the first thing I did was to invite my friends from the Socialist Workers Party to set up an information booth and book display. While at the University of Pittsburgh Mark Ginsburg introduced me to the party, I immediately knew it was an organization I wanted to work with. We would go to any and every picket line in Pennsylvania, Ohio, and West Virginia and walk with the striking workers in solidarity, hold weekly meetings to discuss theory, and I even made my own presentation on Eckardt Kehr, a German historian, and his idea of social imperialism from the 1920s. I housed striking steelworkers from Alabama, helped them raise money for their strike fund, and participated in numerous other activities with the party. I thought it was a natural thing to invite the closest local SWP group to come to Shreveport. They stayed with me while they were in town and after the day on campus they told me the students were very receptive to their ideas and ideals. The SWP is a Trotskyite party and I too believe in perpetual revolution. The next day when the SWP members left to go back to Houston, I was called into my chair's office. She said I reminded her of Joe Kincheloe when he was on campus in the 1980s, and I did not want to have the label of campus radical. I said, I don't? That first semester colleagues would approach me and proclaim that I reminded them of Joe Kincheloe. Soon, I could tell by the tone of their voice whether they liked Joe or not. Shreveport is a conservative city so most of the professors did not like him. Joe wore this as a badge of honor since most of the people in the college of education in Shreveport were anti-intellectuals. It is a strange thing how universities are labeled liberal or radical but most faculty are instinctively conservative. When people would make these remarks I would say to myself who is Joe Kincheloe. I thought if I am going to be Joe Kincheloe I better read his work. My life has not been the same. I am honored that I reminded people of Joe.

It has been nine years since Joe died of a heart attack and there is hardly a day that passes that I do not think of him and thank him for all Shirley and he have done for me and so many other scholars. Imagine how sparse curriculum studies and educational foundations would be without their book series with numerous publishers including this one! After I started reading Joe's work like *Getting Beyond the Facts* (the book that took many years for Joe to find a publisher and transformed Peter Lang into a major education publisher) and other early works. I had to meet this guy and discuss an idea I had for an edited book which became Toby Daspit's and my work *Popular Culture and Critical Pedagogy*. I went to the American Educational Studies Association in Chapel Hill in 1995. I waited outside the conference room where he was pitching Shirley's and his idea that became *Kinderculture*. When the session was over I introduced myself and we bonded over our stories about LSU-Shreveport. What was clear to me then and is still clear today, Joe loved life, books, and democracy. All were connected.

This next chapter is the linchpin to the whole book. I discuss the importance of challenging neoliberalism while creating an active democracy in which people are involved in science matters. I focus much of my attention on becoming an "expert" in the sciences without becoming a scientist. I obviously am convinced that if curriculum scholars are to be a part of a movement to save our dying USA democracy, it will require us to become more involved in scientific matters. It certainly is not the only way we need to become involved, but we have to make sure science and democracy remain connected in a vibrant and meaningful manner. While previous generations just assumed the two were naturally connected. It is clear with the rise of neoliberalism nothing can be taken for granted or assumed; everything must be fought for including the idea of a living democracy. Joe was a fighter and he was interested in anything that was a matter of social justice, equality, freedom, and a living democracy. The best way we can honor Joe is to continue his struggle and fight to create a strong democratic society not just in the USA but the whole world. I know Shirley continues to raise her voice in the name of radical love and democracy. Her voice demonstrates that she still shares an indivisible bond with Joe. My contribution to Joe's memory is this chapter and it is another way I say thanks Joe! Power to the people and Semper Fidelis my friend.

Chapter 5
The Economics of Science, Neoliberal Thought, and the Loss of Democracy

> Let us restore to the sciences the crush of democracy from which they were supposed to have been protected as they grew.—Bruno Latour (2004, 143)

The cultural critic, Robert Pogue Harrison, who rarely writes about science in his critiques of culture in the United States, does so in his latest work, *Juvenescence* (2014). His voice is added to a growing list of critics who are warning the adults of a living democracy that we are becoming more infantile. There is a difference between being child-like and childish. To be childish is to participate in character assassinations, choose shouting matches over reasoned debate, and act more like adolescents than adults. To be childlike is to have a sense of wonder, to see the world anew, and to raise important, critical questions about life. For Harrison, science is one area where the childlike can be nurtured. "Modern science cultivates amnesia," Harrison (2014, 51–52) submits, "so that it can continue to look at the world as if seeing it for the first time. Therein lies its childlike spirit." Now there are two types of memory lose. The first one can lead to a myopia in which the history of science is ignored because the only thing that matters is current knowledge, protocol, and research agendas. The focus is on the here and now, wooing investors, and selling out to the highest bidder. The second form of memory lose is what interests Harrison. This memory is more of a suspension than a loss. In order to see the world anew it is important to ignore the constraint of protocol without losing the understanding of why a protocol was created in the first place; it is to suspend tradition without losing connections to the past and future. A childlike memory asks why this way and why not a different way; why is something true and cannot it be true another way? Few ever think this way mainly because few ever have the opportunity to do so. As with anything that is new, there is a risk that one's childlike thinking might result in a scientist being ostracized from a discipline; ridiculed and mocked as irresponsible. However, there are plenty of examples of scientists thinking like a child and eventually changing how a field of knowledge is conceptualized and operationalized.

The same can be said about a democracy. In order for a government run by the people to thrive, it must have a childlike spirit in which each generation defines

© Springer International Publishing AG, part of Springer Nature 2018   95
J. A. Weaver, *Science, Democracy, and Curriculum Studies*, Critical Studies of Education 8, https://doi.org/10.1007/978-3-319-93840-0_5

democracy on its own terms. Call it a Trotskeyean notion of perpetual democratic revolution. The risk is the new conceptualization of what a democracy is may fail, may place democracy at risk, and cause generational and inter-generational conflict. Revolution can lead to something other than democracy. The reward is democracy may become more vibrant than ever before as more people, not corporate entities, enjoy the benefits of living free and equal. The problem is too many USA citizens are losing their rights, a plutocracy is emerging, politics is becoming more theatrics and circus than deliberative, and adults are becoming childish in their behavior.

As I mentioned in the previous chapter, science and democracy have been connected since the end of World War II by such scholars as Karl Popper, Friedrich Hayek, Michael Polanyi, and most famously Thomas Kuhn. The problem is because of the rise of neoliberal thought, as promoted and developed by intellectuals such as Hayek and Polanyi, and neoclassical economics the assumed natural connection between science and democracy is experiencing a decoupling in which science is more and more connected to multinational corporate agendas than nation-state, democratic agendas. In a neoliberal order the will of the people, if consulted at all, is strictly defined as the needs of consumers or clients and these needs are often strictly defined by multinational corporations. In this chapter, I will discuss what neoliberal thought is, why it is dangerous to a living democracy by, of, and for (all) the people, the role neoclassical economic thought plays in the rise of neoliberal dominance in the world, and how the economics of science offers a strong critique to challenge both neoliberal thought and neoclassical economics. As a result I want to focus on how neoliberal and neoclassical orthodoxy threatens the vitality of the university and curriculum studies. Part of the problem of the decoupling of science and democracy, is the so-called advocates of firmly connecting science and democracy, such as Kuhn, Hayek, Polanyi, and even contemporary thinkers like Helga Nowotny and D. Wade Hands, do not trust citizens within a democracy. Non-scientific individuals involved in science matters and policy make these scholars nervous and anxious. Instead what is offered by these advocates is a bond of trust that citizens of a democracy should just believe what scientists will do and are doing is best and right for a thriving democracy. With the decoupling of science from any concept of democracy, we begin to see that there is a need to critique these advocates and reconnect the two in a manner that strengthens democracy and involves more people in science policy decisions—the very idea that Kuhn and Polanyi, especially, deeply feared.

5.1 The Neoliberal Thought Collective

Neoliberal thought is often explained and rightfully excoriated in curriculum studies and other education literature but rarely are the roots of this international movement explored and explained. Neoliberal thought grew out of specific political and economic experiences prior to and during World War II in Europe and eventually grew most prominently in the United States after 1980. There are numerous dimensions of the Neoliberal Thought Collective that are fundamental to its philosophy but deeply paradoxical and completely toxic to the formation of democratic societies in spite of its proclamations lauding the importance of freedom. Neoliberal thought is a stealth movement originally committed to the undermining of Communism and Fascism and now committed to the destruction of anything public except when public entities can be used to prop up, maintain, and buttress private corporate interests. Often the term neoliberal is used by critics to describe conservative supporters of liberalism, but in early meetings to create a thought collective a few members had no problem with using the term. Many conservatives in the United States object to the term because it has connections with liberalism which they confuse with progressive thought. This objection is a childish response that reflects a lack of historical understanding of economics and political thought in Europe and United States in which liberal and conservative are not as far apart as conservatives today in the United States would like to believe and the term neoliberal is more of an economic term used to describe an extreme, right wing political philosophy after World War II. In regards to economics, the term liberal was originally used to describe those political economists who favored "free" markets. By the late nineteenth century it was a term used to mark conservatives who favored a laissez faire approach over anything resembling socialism or communism. The reference to being neoliberal is an attempt to recapture this liberal economic approach from Adam Smith and David Ricardo with some major addendums that will be explained later to highlight why the emphasis is on the neo rather than liberal when discussing the difference between classical and neoclassical economics.

Although there are numerous economists and political philosophers who can be highlighted as founding fathers of neoliberalism including Ludwig von Mises, James Buchanan, Fritz Machlup, and Gottfried Haberler, the two major figures in the development of a neoliberal movement are Friedrich Hayek and Milton Friedman both eventually connected to the Chicago School of Economics, although Hayek's role in developing this school is limited. Hayek is most famous for writing *The Road to Serfdom* which became an immediate success in conservative ranks when it was published in 1944. It was Hayek's warning of what is to come if science and government succumb to centralized planning. It would mark the end of freedom of thought and only a free market could guarantee that freedom of thought and economic exchange would flourish. Of course, now 70 years after its publication and seeing firsthand the outcomes of neoliberal policies, we can say that the road to serfdom leads to an United States form of fascism in which the market promotes freedom for the politically connected and independently wealthy while serving as a dictatorial system for the majority who are denied union rights, legal rights, and equal access to policy makers. This course of limiting access to political and economic power for most people in free societies from the very beginning of neoliberalism cannot be construed as an accident. Neoliberalism always was a threat to democracies.

The Neoliberal Thought Collective that Hayek and a few others envisioned is also referred to as the Mont Pèlerin Society and the philosophy undergirding neoliberalism reveals everything about the tragic consequences of this movement. In his own reflections on a neoliberal thought collective Hayek noted "it must remain a closed society, not open to all and sundry" (Plehwe 2009, 16). In the name of freedom, thought had to be controlled, access limited, and only true believers admitted. Like a vanguard of the proletariat or a fascist special unit, the Mont Pèlerin Society was to serve as the inner sanctum where right minded thinkers "who shared philosophical ideas and political ideas could mingle and engage in the a process of further education and collective learning dedicated to advancing a common neoliberal cause" (Plehwe, 5). This meant from the start neoliberalism believed the only way to guarantee the "right" form of freedom was to force it upon the people of the world. This remains an important attribute of neoliberal thought in which neoliberals are encouraged to think freely while denying anyone else the right to do so. This meant that as part of its philosophy, neoliberalism would have to simultaneously use political force to ensure the establishment of neoliberal principles while publically denying that political action is ever warranted and necessary for anyone to use since only "the market" knows what is best for society in general and no individual could know what "the market" knows. As a result and from the start, the Neoliberal Thought Collective rejected the idea of any organized action against "the market" including trade unions and monopolies. By the 1970s, however, The Mont Pèlerin Society continued to despise labor unions, but learned to love any form of corporate organization and most certainly monopolies. This love affair with corporations continues to this day and has led to the creation of such corrosive organizations such as the American Legislative Executive Committee who sponsor many anti-union bills in state houses, school to prison pipeline policies, and stand your ground bills passed into law including in Florida. It has also taken the lead in defining corporations as individuals and encouraged unlimited campaign contributions leading eventually to the Supreme Court ruling in the now infamous Citizens United v. the Federal Election Commission case in 2010. However, this same power flexed by the Neoliberal Thought Collective and corporate organizations cannot be shared or allowed in a neoliberal order. As the economist of science, Philip Mirowski (2013, 56) notes "one implication" of the denial of the importance of political action "is that democracy, ambivalently endorsed as the appropriate state framework for an ideal market, must in any case be kept relatively impotent, so that citizen initiatives rarely are able to change much of anything." Recent evidence of the consequences of this neoliberal tenet is citizen movements to raise state minimum wages through ballot initiatives only to see state senate and house Republican representatives attempt to undermine these initiatives or corporations such as McDonalds to use the courts to challenge the will of the people.

The Neoliberal Thought Collective's movement to simultaneously use political power to entrench their own agenda while denying those same rights to citizens of a democracy is founded in a firm principle that proclaims the "free market" decides everything as the ultimate arbitrator but just in case the market does not political force must be used to make sure "the market" and all political structures rule in favor of neoliberalism. Neoliberals proclaim that if given a choice citizens of a democracy would choose the solutions to problems constructed by "the free market" but just in case they do not neoliberalism ensures that there is no other option such as community organization, collective bargaining, social and political protests, or other communal means available in "the free market" for individuals to select. In the name of freedom, therefore, neoliberalism proclaims that there is only one choice and that choice, if need be, will be made willingly or coercively through courts of law, police forces, and of course in the case of the neoliberalism's grand experiment, the Pinchot government, military coups. A consequence of this neoliberal plank is the separation of freedom from democracy. "Neoliberals extol 'freedom' as triumphing over all other virtues," Mirowski (2013, 60) notes, "but the definition of freedom is recoded and heavily edited within their framework." Apparently the "free market" is only free to affirm neoliberal core beliefs and only neoliberals are wise enough to manipulate human political structures and the ubiquitous market in order to construct a society. Such a philosophy sounds less free and democratic and more tyrannical with every new development in politics throughout the world. The next financial crises (caused by neoliberal policies) will bring an even more forceful assertion of neoliberal doctrine.

To overcome the deep flaws embedded in the foundation of neoliberal thought there is a need to foster doubt and ignorance. The historian of medicine, Robert Proctor (2008) has called the study of this rise of doubt and ignorance agnotology and the political theorist Wendy Brown (2015, 189) refers to it as a "smugness in ignorance". I want to focus on the role economists play in fostering doubt and ignorance in the name of the Neoliberal Thought Collective. Philippe Pignarre and Isabelle Stengers (2011) use the term minions, and this is exactly what economists who have peddled neoliberal thought are. However, as Pignarre and Stengers (2011, 33) note "naming minions is dangerous." Indeed it is. Calling economists minions does not mean I am referring to all economists or even all neoclassical economists because like any field of knowledge economics is diverse. The term neoclassical is as broad a term found in economics. The term is often used to name progressive economists like Nobel Prize winner and co-creator of the General Theory of Equilibrium in economics, Kenneth Arrow who is in favor of government action in markets when the markets fail. Neoclassical can also be used to describe Milton Friedman who is one of the founding members of the Mont Pèlerin Society and its most forceful advocate. Ironically, and to demonstrate the diversity of this term, one cannot call Hayek a neoclassical economist if one uses mathematics as a major demarcation between classical and neoclassical economics since Hayek was not well versed in mathematics nor very well accepted by neoclassical economists. When economists are described as minions of neoliberal thought it also does not mean they are unthinking drones merely following the orders of a mastermind. Minions are highly specialized intellectuals who willfully use obfuscation and doubt to structure policies to favor one group over another. Minions use reason to rationalize a course of action that naturalizes the willful cause of misery and suffering of billions of people and rationalize that certain policy outcomes are inevitable and out of anyone's control. "There has…been a name for something that manages to produce a coincidence between enslavement, the putting into service, and subjection, the production of those who do freely what they are meant to do….Its name is sorcery" (Pignarre and Stengers 2011, 35). Neoliberal economists' sorcery comes in the form of denying there is any alternative to the "free market," and organized resistance to the "free market" is futile because the system is too colossal to control, explain, or understand even for highly trained economists. Economic sorcery has become a form of mysticism.

Economists as minions play other roles as well. They also serve as noise machines that place in doubt any alternatives emerging from public discourse as to what causes economic calamities. As Mirowski (2013, 297) suggests in *Never Let a Serious Crisis go to Waste* economists are on the airwaves, in the chairs testifying at the federal and state levels, and at the shareholder meetings "to pump excess noise into the public discussion of appropriate frames within which to approach the controversy." This means the economists are there to dole out economic and political blame and more important they are there to express doubts that any state or public sanctioned reaction against "the market" can make a difference and assuring that no legitimate alternative to a neoliberal "free market" ideology can emerge. If there is an attempt to alter the "free market" economic minions are expected to simultaneously offer up solutions to a current economic crisis and, then, when anything happens, to declare the solution a failure and ensure the "free market" will return to its normal freewheeling ways. This is exactly what is happening after the market collapse of 2008. Soon after congress imposed regulations curtailing banks and the financial sector attempts were made by American Bankers Association, headed by Frank Keating former Governor of Oklahoma to undermine the new regulations citing they were stifling competition and undermining the "free market." From the collapse of 2008 to 2011 after Republicans took over the House of Representatives very little was learned as the root causes of economic crisis were covered over by obfuscation and doubt while the minions were in place to make sure nothing would succeed in limiting the "free market." As of 2018 the economic noise machine is still in place to assure, after the next financial crisis comes, no one on Wall Street will face any prosecution, no corporation will be held accountable, the "free market" will remain intact, and, most importantly, when the time comes appropriate scapegoats are found to blame. As I write this, reports are beginning to emerge of a planned Federal bailout of overextended oil companies as the price of oil declines rapidly in an attempt by larger oil companies to eliminate smaller competitors. True to neoliberal doctrine, federal money will be used to feed oil companies to insure they will not lose money and then this crisis will be used to declare that food stamps and Medicaid are entitlements that cannot be sustained and must be cut. True to their script neoliberals have instituted a massive tax cut for the wealthiest USA citizens, denying the federal government a source of tax revenue, and just 2 months after passing these tax cuts neoliberals minions have proclaimed that the federal debt is growing because of entitlement programs like medicare and social security. What is worse the minions who deceive a democracy or the sheep who allow it to happen?

Wendy Brown refers to this as shared sacrifice. Since it cannot be expected that corporations, the heart and soul of any heartless and soulless neoliberal system, will risk the loss of profits, someone else will have to sacrifice for the good of the neoliberal order. As corporations are enjoying the benefits of slack regulations, tax benefits, cuts, breaks, and abatements, and a constant flow of taxpayer funds in the form of government subsidies or contracts, when economic crisis emerges another sector of society has to sacrifice. Wendy Brown (2015, 210–211) notes shared sacrifice takes the form of "austerity" measures, "sudden job losses [with no assumption of unemployment benefits since the individual entrepreneur is at fault for any job loss], furloughs, cuts in pay" and more importantly in the long term "curtailed state investments in education, infrastructure, public transportation, public parks, or public services."

Another important dimension of the Neoliberal Thought Collective is reduction of all life to Homo Economicus and one way this is accomplished is to view knowledge as a commodity. Wendy Brown (2015, 30) is the latest to explain the dominance of Homo Economicus. She writes: "I join Michel Foucault and others in conceiving neoliberalism as an order of normative reason that, when it becomes ascendant, takes shape as a governing rationality extending a specific formulation of economic values, practices, and metrics to every dimension of life." Homo Economicus in neoliberal thought dominates everything, including that which cannot be explained by an economic rationality. Learning is no longer a life changing experience meant to build character or expose young minds to different ways of seeing the world; learning is now determined by return of investment and defined as excellence based on the employment it can offer after one's education is complete. Teaching is no longer a field defined by a passion for inquiry, a desire to mold the young, or a curiosity to seek out answers to problems, instead it is job training and teacher efficacy is defined by test scores and customer satisfaction. Homo Economicus creates a frenzy in which everything in life has to have a utility. There is no time for leisure or thinking unless it can be construed as a necessary path toward an economic gain. The only acceptable activity outside of work is escapist experiences such as sporting events, parties, and perhaps vacations, but these events are often construed as opportunities to be seen and network. What emerges from Homo Economicus is a new individual who is an entrepreneur, a free agent and "human capital." There is no need for social organizations, certainly no need for political organizations because everyone now is their own corporation. Demanding a livable wage, benefits to support a family, and a pension are now unreasonable demands. An entrepreneur does not make these demands on a corporation, instead they go and get these life rewards themselves. Again Brown (2015, 80) along with Foucault is insightful here, "neoliberal man comes to the market, as Foucault puts it, 'being for himself his own capital, his own producer, the source of his earnings.' Whether he is selling, making, or consuming, he is investing in himself and producing his own satisfaction." Homo Economicus does not complain about injustices or inequalities, these antiquated concepts are no longer valued. The only victims in society are lazy individuals unable to invest in and sell themselves as capital. To invest in them as a nation-state would waste valuable investment funds that could go to more important areas such as tax breaks to bring corporations from one state to another. It is the responsibility of the individual to serve as their own welfare state while the nation-state at all levels is now responsible to the corporation, making sure they are fully supported including presented with an abundant number of docile workers who have been properly trained to expect nothing from government and corporate institutions.

Homo Economicus replaces all other forms of life including Homo Politicus. Homo Politicus comes from the Aristotelian notion of the good life in which through reason, leisurely thought, and political involvement people are able to hold at bay the "'unnatural'" tendency to "incite the desire for wealth for its own sake" (Brown 2015, 90). For Aristotle, the very existence of Homo Economicus should not be allowed to emerge in a democratic society. Wendy Brown demonstrates that throughout Western political thought Homo Economicus was thought to be unhealthy and unnatural. As she points out even Adam Smith believed that Homo Economicus should be tempered and controlled by "deliberation, self-direction, and restraint, all basic ingredients for sovereignty (Brown 2015, 93). Smith was not alone in sharing these Aristotelian sentiments. Rousseau also believed that "Homo Politicus…must literally subdue the creature of self-interest and self-absorption" (Brown 2015, 95). John Stuart Mill believed "that political economy 'makes entire abstraction of every other human passion or motive' and thus operates with a fictional subject" (Brown 2015, 97). Smith and Mill were both major classical economic writers and even they did not trust Homo Economicus as a major form of life. Yet neoliberal thought has ignored 200 years of political and economic thought. As Brown (2015, 92) notes "Aristotle has been inverted, if not buried." In place of Aristotle and the tradition that followed, neoliberals erected Homo Economicus and there this fictional character sits on the altar of human existence worshipped by the powerful and self-interested.

Homo Economicus became what Foucault referred to as "Market Veridiction." This meant "the market was a new site…for governing…it produced, organized, managed, and consumed individual freedom, all without touching the subject" (Brown 2015, 58). Homo Economicus became the arbiter of truth, replacing gods, kings, and people while never banning a holy book, firing a shot, or forcefully coercing a single person. Instead what Homo Economicus has created is a dependency on the market and the market spokespersons who take the form of corporations, think tanks, economists, Supreme Court justices, politicians, and select academics. Any attempt to proclaim an injustice and organize against Homo Economicus is both unnatural and untenable since free agents are not only on their own, but they do not need any help.

One final aspect of the Neoliberal Thought Collective is its drive to convert knowledge into a commodity. Traditionally knowledge is viewed as something that often evades definition because it defines a wide variety of ideas, concepts, outcomes, tangible objects, practical creations, inventions, artistic endeavors, folktales, common sense, myths, poetry, genetic material passed from generation to generation, and other possible forms as well. In a traditional sense knowledge is often construed as both practical or useful and useless. It is also defined as something that leads to a career and character development. With the rise of neoliberal thought and Homo Economicus, knowledge is now defined as something that is a commodity that can be patented and eventually either sold or turned into something that can be sold. As a result there has been a rush to patent almost anything that is created in a laboratory or university setting and to construct technology transfer agreements in which one university will agree to share a certain research protocol or invention with another university if the second university will agree to share any possible future profits. Like everyone else toiling under the tyrannical banner of Homo Economicus, professors and university researchers now are expected to be knowledge entrepreneurs. "Neoliberalism preaches," Mirowski (2011, 34) suggests, "that one must actively construct an ideal market, not just wait for it to appear on its own." The entrepreneurialization of the university has led to major changes in the university. It has created a logic of usury rather than one of intellectual community building. As an entity of community building, the university is a holy place in which young people come to discover who they might become, including something they never dreamt possible, and what their roles in life and society might become. It is a place where people filled with curiosity and wonder explore research problems not because they offer the most potential for profits and grant money, but simply because someone is curious and wonders if there is a solution to a problem. These problems, in a university of community building, can be practical problems or they can be the most obscure and obtuse imaginable. For instance, when Newton fled London for the pastoral life in Cambridge he did so to avoid the outbreak of a plague. Had he stayed in London he could very well have discovered the cause and a cure for the plague. Instead he chose the very impractical and developed a form of calculus and chose to stick is his head in the celestial stars, way above the clouds.

Counter to the university as a community is the university as a form of usury. The usury university is a place where interactions are calculations, chance encounters are avoided, elective courses are wastes of time, course work is resume building, and intellectual relationships are based on what material goods each can give the other. The usury university is an unhealthy institution and this is why it is viewed by neoliberals as the ideal form of learning.

The consequences of the neoliberal usury university are severe. In 1802 Friedrich Schelling gave a series of lectures that eventually became a book titled *On University Studies* (1805/1966). In these lectures he said the following about the importance of teaching and research: "A man who lives with his science as though on another's property, who is not himself in possession, who has never acquired a sure and living feeling for it is, who is incapable of sitting down and reconstructing it for himself, is an unworthy teacher even when he attempts no more than to expound ideas of others (Schelling 1805/1966, 26–27). This is exactly what is happening in neoliberal state universities. Because of the principle of "shared sacrifice" one of the institutions sacrificed is a fully funded state university system. State universities throughout most states in the union have seen the state portion of funding to cover the cost of educating the next generation decline tremendously in the past 35 years. As a result student loan debt has increased tremendously where all sources report a range of loan debt upon graduation to be on average between 24,000 and 27,000 and on average students also have another 6000 in credit card debt. The so-called student loan debt crisis might be the next financial crisis and given that neoliberal logic will not retreat then we can expect some more shared sacrifice on the part of students, faculty, and universities in order to pay down the debt. Some will point fingers to a dramatic rise in administrative positions at universities, most will blame faculty who barely teach, and students who irresponsibly borrowed too much money. Few, though, will blame predatory lenders, a rigged student loan system that keeps interest rates artificially high, and a neoliberal logic that devalues and defunds higher education for most people but instead pushes technical and community college.

As state universities continue to be defunded students are not the only ones expected to bare the brunt of the cost, faculty and staff are. In order to cut costs, administrators hire fewer tenured, research oriented faculty and instead hire more and more adjunct professors who act as free agents, teaching a class here, another one there, and some more over there. Under this new system these faculty are unable to dedicate anytime to cultivating their knowledge within their field. Instead they become what Schelling called interlopers on another person's property and more importantly unworthy teachers. What faculty are becoming at state universities are hired hands, told to follow a standard syllabus and dole out bits of information that will be useful for the next generation to find a job. What is being created is a two tiered system of elite private universities and liberal arts colleges where faculty are free to explore any field of research that interests them and where federal and private grant funds will flow freely for anyone with a research agenda meeting the needs of corporations and nation-states. These faculty will be world class teachers at the cutting edge of their chosen fields while at the state universities a select few will be able to conduct research, contribute to their field of knowledge, and become worthy teachers. Most faculty at state universities will become interlopers unworthy of the title of teachers since they are disconnected from research. Most alumni will not be bothered by these circumstances as long as they find good paying employment in a career of their chose and the football team is competitive, parents will be concerned with return on investment reports from *US News and World Report* and *The Princeton Review* (both part of the neoliberal rubric), students will care more about the social scene including the reputation of the athletic teams and their future employment prospects then faculty qualifications, and administrators will not care about the disconnect between teaching and research because they only have to throw around the word excellent or real life experience enough times in order to distract the "stakeholders." The disconnection of teaching and research is part of what Philip Mirowski (2011, 195–196) calls the "globalized privatization regime" in which a "vanguard believes that science is really just another manifestation of the generic division of labor, to such an extent that a small elite of captains of cognition can sit atop the entire knowledge economy pyramid and direct the serried ranks of worker bees."

These circumstances do not only apply to state university faculty. Worker bees are also found in corporate laboratories that are more and more being farmed out to specialized corporations that are seeking cheap labor to do the everyday tasks required to conduct research in a laboratory. This neoliberal model of commodifying knowledge has led to uncertainty for laboratory scientists in all fields of knowledge and institutional types. In a debate with Bruno Latour over the state of knowledge in the field of sociology, Steven Fuller (Barron 2003, 97) highlighted the consequences of these circumstances. "To be sure, we are living in a period where it's convenient to forget the moral aspect of science because researchers are not institutionally protected any more. Research is done in many different contexts, the terms of which we often cannot dictate. From that standpoint, we have to make the most of what we can out of the situation…If you're a researcher without much standing on your own, as many fixed-term researchers are today, then you've got to be able to satisfy your clients." While I think Fuller is off target in suggesting that it is Bruno Latour's anthropological approach to science that makes such circumstances possible or more palatable for researchers, Fuller is suggesting that quality research is compromised. There are plenty of examples to draw from the pharmaceutical research area alone to suggest Fuller is quite correct.

The neoliberal university creates a starved culture in which the millennial old quest for the good life is being replaced by a quest for mere life. In the good life quest alternative paths are explored, ideas tested and debated, characters formed, and personalities shaped for a lifetime. In a quest for a mere life, college careers entail luck. The student should feel lucky that they are able to get a college education and they better not waste their luck on some meaningless degree like philosophy, history, or literature. Those degrees are for the leisure class who attend the elite schools. The good life is contemplative. Mere life is calculated, direct, and focused with no time for detours just a linear line from freshmen classes to sustainable employment at least for now. The good life is being educated by experts who dedicated their lives to learning and creating knowledge. Mere life is an education based on skills, outcomes, objectives, and other neoliberal metrics that lead not to a degree with meaning but a certification that serves more as a tattoo that can be shown to others rather than a developed mind seeing the world differently.

Dominique Pestre assessment of the Neoliberal Thought Collective affirms Brown's insights. For Pestre (2012, 244) neoliberalism seeks to "reassert the individual as alone in responsibility for his/her own life after the unfortunate digression into Welfare Statism and Keynesian economics." Pestre's prognosis to the ills of the individual within a neoliberal form of government is only partially exposed. The Neoliberal Thought Collective wishes to construct a country club style of government in which clear walls enclose those select few who are invited into the government while it is Hobbesian anarchy on the outside. Since the individual is responsible for oneself and there is no safety net each person enters into an anarchical arena to secure their own well-being, creating clear demarcations between winners who live a life of temporary security teetering on the brink of collapse and losers who are lazy and undeserving of enjoying any security. While this anarchical street fight is taking place, inside the walls of the country club neoliberal government, welfare statism is alive and in a generous mood. Rich members and corporations are welfare recipients of lush gifts funded by taxpayers who are relegated to live outside the country club walls but obligated to fund "government" programs. Corporations are guaranteed profits via tax abatements, tax credits, even a zero tax obligation, and outright gifts from the neoliberal welfare form of government. The rich individuals, especially those born into wealth, also enjoy the perks of the country club welfare state.

Of course, in reality the vast majority of people in any form of democracy would not accept such conditions so the transition from representative government to the road to neoliberal serfdom has to be covertly established through "a new form of social engineering" (Pestre 2012, 246) "remaking people and institutions, by reformatting their behavior" (Pestre 2012, 245). This social engineering occurs just as it does for creating a country club form of government. Welfare is not for the people, it breeds laziness and contempt while the welfare state remains alive and thriving for the wealthy and the new citizens of the Neoliberal Thought Collective, corporations. In the social engineering efforts of the Neoliberal Thought Collective, who have in the past renounced all efforts of social engineering, "there is no need for institutions whose function is to define 'common interests' (Pestre 2012, 246). Are you dissatisfied with society and the political order? Too bad there is no need or hope in collectively protesting. Do you feel discriminated against because of your social class, religion, race, gender, or sexual orientation? Too bad there is no institution for you to complain to because there are no African-Americans, gays and lesbians, women, atheists, Muslims, or working class groups just individuals who need to negotiate everything they need with other individuals. The neoliberal country club government is not here to help you. They are too busy serving the truly needy, the job creators and corporations who need more wealth, more land, more cars, more homes, more laws supporting their interests, more lobbyists, and more representatives in all forms of government. It is tough being an individual and a corporation at the same time in this global economy. They need help, the rest of you are lucky to have your guns and abortions, now there are freedoms you can trust. One should always have a gun and access to abortion clinics when one is homeless and destitute. Once again in the Neoliberal Thought Collective world, freedom wins, although most cannot afford to be free.

One final consequence from the commodification of knowledge is the end of the market place of ideas. The Neoliberal Thought Collectives goal was never to create a marketplace of ideas where Derrida's (2004, 148) vision of a university prevails and is "open to types of research that are not perceived as legitimate today, or that are insufficiently developed…including some research that could be called 'basic.' We should go one step further, providing a place to work on the value and meaning of the basic, the fundamental, on its opposition to end orientation." For sure Derrida would never have been invited to a neoliberal university. He thought too much and was too open minded to the creation of new possibilities for seeing the world in a childlike manner. The neoliberal university is too childish for Derrida. The neoliberal ideal of a market place of ideas is less of an emphasis on ideas and more of a focus on those ideas that can quickly be converted into marketable goods. In this sense the university open to a market place of ideas was always a sham. The history of universities always were dedicated to and controlled by (mostly) men who were interested in money. Philip Mirowski (2011, 35) notes that the neoliberal notion of a free market of ideas is "geared to submit all ideas to the refining fire of dollar votes, within a consciously structured interlocking set of economic markets." The difference between universities fifty years ago and the neoliberal university today is not that one is interested in practical research and the other was interested in high ideals. The difference is in degree. The university, especially state universities are open for business more than they ever have been and the abandonment of the quest of knowledge for its own sake is no longer even considered an important attribute of what it means to be a professor or a graduate of an institution of higher learning. A rubric could not be created to quantify this ideal so the Neoliberal Thought Collective pushed it out the door.

5.1.1   Decoupling Science from Democracy

The rise of the neoliberal university and the consequences of this dominate ideology in managing universities has created a serious test to the strength of the link connecting science and democracy. In his next section I want to show that the connection between science and democracy has been always weak because scientists and those studying scientists for the most part are wary of democratic involvement in science policy decisions. In fact people such as Hayek and Polanyi were openly hostile to the idea of a democracy and advised scientists to embrace the mystical free market as an alternative to any form of democracy. When the connection between science and democracy were broached democracy is usually defined as an internally operational definition functioning within the "Republic of Science" and since scientists are by nature democrats, citizens need not worry about the decisions scientists were making or the research they were conducting, All was done in the name of democracy. With the rise of a neoliberal ideology, this connection between science and democracy is no longer made. As a result, there is more need to push for citizen involvement in science policy decisions. This will require major educational changes and more effort of citizens to become experts in a chosen field of science.

5.2   The Angst of Autonomy

In his seminal work on the economics of science D. Wade Hands discusses Partha Dasgupta and Paul David's latest attempt to defend the autonomy of science. Hands (2001, 377) writes the "political message that emerges from Dasgupta and David's new economics of science is the defense of the autonomy of (open) science. The ability of science to produce and disclose reliable knowledge is highly sensitive to changes in the underlying reward structure of science." Government funding of course should flow freely but these financial flows "should not come with governmental strings attached" (Hands 2001, 377). In other words, dictate too much as to what scientists should and should not research, and limit their funds too much, the talented scientists may lose all incentive to do research. However, if, under the right conditions, the scientist is left alone, they will produce reliable and publically valued knowledge. This angst over the loss of scientific autonomy is not an uncommon thread in the writings of philosophers and historians of science or of scientists themselves. This tradition can be traced back to at least Immanuel Kant and his book *Conflict of the Faculties* (1798/1979) and is a theme that appears in Edmund Husserl's *The Crisis of European Sciences* (1954/1970), the current angst over the autonomy of science emerges from the interwar period in Europe and during World War II. Hayek and Michael Polanyi during the war were the leading advocates for an autonomous science community and both introduce the "market" as the solution to prevent disastrous government (central planners) intrusion, misguided philosophy (positivism), and over generalized knowledge (statistics and data) from corrupting the scientific process. Another theme that undergirds most of the anxiety over the loss of scientific autonomy is the fear of relativism which is another way of saying the fear that democratic forces including the will and opinion of "the people" will shape the nature of scientific work, denying science its uniqueness. All of these "threats" were growing in popularity and importance throughout the world and Hayek and Polanyi were warning scientists about these imminent threats.

Anxiety for Hayek took many forms including the rise of centralized planning not only in the dictatorial Soviet Union and Fascist Germany but also in European Socialist nations including his home in exile Great Britain. Centralized planning, coupled with the rise of positivism in which perfect knowledge could be achieved, and statistics from which general laws and policies could be deduced, was constructed on a false premise that one person or a group of people could master all the knowledge that was needed to make rational economic decisions. Central planning, positivism, and statistics, "cannot take direct account of these circumstances of time and place," Hayek (1948/1980, 83) insisted. Economic exchanges are rapidly changing and dependent on factors that no one individual could control. As soon as central planning would make a decision based on available information it would already be obsolete because "the economic problem of society is mainly one of rapid adaptation to changes in particular circumstances of time and place" (Hayek 1948/1980, 83). The only solution was to let decentralized planning take its "natural" course and that competition or the independent gathering of information by all individuals in a society was the only means to insure the development of reliable scientific knowledge that could be used to make rational decisions. It was the "market," naturally decentralized and competitive that would produce the best conditions for a society to flourish.

Michael Polanyi worked from Hayek's ideas about the "market" as a decentralized generator of knowledge and applied it specifically to science. By their nature, scientists operated in a "Republic of Science" in which they were free to conduct the research they believed to be most important and pressing for a society. The only supervision these republicans of truth needed was their own community of scholars who would make sure that the correct research was conducted under the guidelines of the correct protocol. This is what Polanyi (1969/2002, 467) referred to as "self-coordination" and "their co-ordination is guided...by an 'invisible hand' towards the joint discovery of a hidden system of things." Any interference with the mystical functioning of the invisible hand "to organize the group of helpers under a single authority would eliminate their independent initiatives and thus reduce their joint effectiveness" and "paralyze their cooperation" (Polanyi 1969/2002, 467). This interference, it is assumed, not only includes government interference but also any industrial attempts to create a monopoly. The best place for the Republic of Science to exist is in the universities. And in a moment of democratic spirit Polanyi rationalizes this conclusion by defending "the public". "Those who think," Polanyi (1969/2002, 479) warns, "that the public is interested in science only as a source of wealth and power are gravely misjudging the situation." "The public" would be willing to support science for the sake of exploration and wonder. Therefore, scientists should appeal for support to pursue scientific endeavors in the universities because "the universities provide an intimate communion for the formation of scientific opinion, free from corrupting intrusions and distractions" (1969/2002, 479). Then the Platonic Polanyi (1969/2002, 479) returns when he reminds his readers that once science is firmly established in their communes of higher learning with public support, "the general public cannot participate in the intellectual milieu in which discoveries are made." The Republic of Science was democratic internally as all scientific voices were welcome and needed in order to create the best truth seeking protocols and deciding on the most important and fruitful research agendas, but when it came to intruders from the outside the scientific community returned to its Burkean roots in which "It rejects the dream of a society in which all will labour for a common purpose, determined by the will of the people" (Polanyi 1969/2002, 485).

The reluctance to include "the people" in scientific endeavors is still a common theme today. It can be found in Hands work on the economics of science. Hands is less restrictive than Polanyi but still very reluctant to open the gates of the Republic of Science. Like the others, Hands is not willing to abandon the uniqueness thesis of science. This thesis recognizes that scientific communities are social constructs and no different than religious, cultural, political, and economic entities in this regard. Yet, those who advocate this thesis are quick to point out that science is still unique from these other social structures. Science may be influenced by belief systems like religions are, steeped in politics like any other organization, shaped by cultural values, and involved in economic exchanges like corporations and individuals are, but science is different. Its values, beliefs, and conduct are like no others, therefore no matter how alike science may be to other organizations it needs and deserves its special autonomy from any restraints. Hands (2001, 401) accepts the "idea that science is social." However, like the others mentioned he is reluctant to push this notion outside of the parameters of the scientific. Hands recognizes that there is an importance to ask "the question of whose interest is served" when discussing economic methodologies and scientific knowledge, and he aptly quotes Steven Shapin and Simon Shaffer's *Leviathan and the Air Pump* where they write "'Solutions to the problem of knowledge are solutions to the problem of the social order'" (Hands 2001, 402). Yet, when the opportunity to pry open the gates of the Republic of Science, the furthest Hands (2001, 376) is willing to go is to recognize that "the system of open science does not exhaust the totality of the 'scientific community.' In addition to collegiate science, there exists a parallel system of applied science and technology that is housed primarily in research facilities of specific industries and the military sector." This admission alone is reason for the gates of the Republic of Science to be opened for public involvement and scrutiny. It is no revelation that science has been since the beginning of time connected to economic and military interests, but it is to recognize that every single individual has a right and duty to explore these connections and demand a voice in determining how corporations will use science and how the state will use science for military purposes. Although corporations may function under the guise of a "free market," this market should never be closed off from public scrutiny. Corporate research should not be conducted in secret when the consequences have deep public safety concerns. The "free market" and certainly nation-state laws are used when it comes to corporate interests to keep the public in the dark and certainly on the outside of the gates of the Republic of Science. This is one way in which the Neoliberal Thought Collective has constructed the individual not as a citizen of a democracy but as a consumer in a corporatocracy in which individuals have no rights except the right to consume whatever corporations produce. These are issues Hands misses when he is reluctant to push open the gates of the Republic of Science and this reluctance continues the long tradition of philosophers and historians of science cloaking the workings of science in a dark cloud of mystery rather than shedding light on the inner workings of science and helping the public understand the importance of science in their lives.

Another well-known scholar concerned with public interaction with/in science is Helga Nowtony. Along with Pestre, Nowotny represents a new wave of science studies scholars thinking about the role between science and democracy. She is concerned with the onslaught of neoliberal thought on science and recognizes the many negative consequences of this thought on a living democracy. "How much public science does society need," Nowotny (2010, 25) asks, "and how public must science be…?" The first question is an important economic and political question that neoliberals often respond in the most restrictive ways when it comes to public policy and the most radical of ways when it comes to corporate interests. For neoliberals the state, which is to say the people other than as consumers, should not play any role in deciding science matters other than in regards to military research and then the experts should provide the answers to any military matters. In others words from a neoliberal perspective there should be no public science. The second question is a philosophical question and the most important question Nowotny raises.

Nowotny believes that science is under assault from two fronts. The first is what she refers to as the "propertization" of science and the second the push to democratize science. The propertization of science is a direct result of the Neoliberal Thought Collective, the attempt to commodify knowledge, and, in my opinion by far, the most threatening issue facing science. Propertization of science covers a whole range of issues. It includes what Nowotny (2010, 2) refers to as the "propertization of scientific data." This can take many different forms including the idea of who owns human data generated from medical tests or the human genome project? Corporations claim proprietary rights over this data since they created the protocol to test and diagnose a patient. Scientific data can also be seen as a growing demand by corporations to receive the data they paid for. In other words, if corporations invest hundreds of millions of dollars in a new drug or a new technique they seem to believe they earned the right to receive the appropriate data to support the marketing of their product. This also means in a growing number of cases that corporations also have the right to hide unfavorable data from the public as well. It also takes the form of conflicts between university researchers who contract with corporations and then the corporations prevent or delay the publishing of dissertations and findings in peer reviewed journals for a period of time delaying the graduation of students and tenure of faculty.

The propertization of science also takes the form of reshaping of the individual. The individual rather than a citizen or an adult living in a democracy in which they have certain inalienable rights are constructed as a consumer who has certain rights but none of them are political or inalienable. When citizens of a democracy are transformed into consumers it becomes their responsibility to seek out the correct and accurate information concerning medical and scientific concerns. If a corporation hides negative data concerning a pharmaceutical drug, it is the responsibility of the consumer to know this occurred and to act the only way it can in a neoliberal world, boycott the drug and select another one. If a medical procedure has potential negative side effects, the consumer is responsible for knowing this. It is not up the corporation to stop producing a product or a medical doctor and scientist to stop utilizing a technique, the consumer has to force action by making wise choices. Nowotny (2010, 3) suggests that this new mentality has "captured the public imagination in the guise of promising greater individual autonomy. The freely choosing consumer is first cousin to the authentic individual." And like any neoliberal idea, this familial relationship is incestuous.

Other dimensions of the propertization of science are common themes already discussed but worth mentioning because they provide more concrete evidence. For Nowotny (2010, 10), propertization marks "the growing influence of new forms of economic rationality." It is a rationalization that is limiting the collective imagination and leading to the dumbing down of societies. As mentioned earlier this rationalization is causing universities, states, and individuals to enter into the legal agreements limiting how certain data and techniques can be used and placing economic claims on any future inventions. This rationalization includes the turning of scientists into entrepreneurs which Nowotny claims places a pressure on scientists to constantly produce something new and marketable. "This process is accelerating—and accelerated by—the dynamics of innovation. Once science could claim to have several, perhaps contradictory functions. Today, its overriding function is to initiate, sustain and be the main driving force behind innovation" (Nowotny 2010, 12).

Like the propertization, the democratization of science takes different shapes. On the negative side, and as a result of a neoliberal ideology that defines the individual as a consumer, the expectations from consumers is that science will deliver them products that will curtail their health ailments and "undesirable side effects and unintentional consequences will be minimized, risks made foreseeable, and if deemed unacceptable, completely eliminated" (Nowotny 2010, 14). This mentality is not only untenable for science to meet, but it constructs the individual as an unthinking, inanimate portal created to deposit the next wonder drugs. The individual becomes lazy; demanding science do their thinking for them, medicine tend to their dormant minds and bodies, and lawyers, if need be, litigate for them if any deception or unmet expectations arise.

Democratization also marks a growing distrust of science which takes at least two different forms. The first form is found in the politics of doubt and ignorance in which evolutionary biologists, molecular biologists, and environmental scientists are discredited not because of the evidence they present to support their theories, but because of the politics of their findings. Creationists and deniers of the Anthropocene seem to believe that science must fit their theology and ideology and if it does not then it is science that must change not their narrow, closed views. The second form includes those who no longer view science as "independent and standing above vested interests" (Nowotny 2010, 2). This can include people who want to know where the funding for a scientific project came from or who is involved in making political decisions as to what scientific projects will be funded and who will conduct the research. As Nowotny (2010, 3) notes this notion of the democratization of science "pushes citizens toward becoming more involved in the priority-setting of the research agenda and therefore in the workings of science as an institution that claims to work for the benefit of society... 'society speaks back to science'" under this model of democratization.

Given this last statement about democratization, I find Nowotny's notion that science is under assault curious. Certainly science is under assault from a populace that is interested more in religious dogma and political and economic ideologies. It is difficult to commune and communicate with the close-minded. How is science assaulted by more individuals wanting to play a bigger role in the decision making process? It is here that Nowotny's view of science is closer to Polanyi's than needs to be I think. Nowotny (2010, 5) acknowledges that science "is public and depends on having a public" but she quickly closes the gates when she adds "which is first and foremost itself." It is not until something is published that "scientific knowledge becomes public and accessible in the public domain," then comes another caveat, "even if it can only be fully understood by other specialists" (Nowotny 2010, 5). She adds the "peer group remains the only arbiter believed to be qualified and sufficiently trusted in assessing the production of reliable scientific knowledge" (Nowotny 2010, 6) There is no doubt that scientific knowledge is constructed via a public that starts with itself as specialists and peer review is essential. It is what currently distinguishes solid evolutionary biology research and creationist religious doctrine confusing itself as science. The scientific public need not stop with specialists and peer groups, this is merely a continuation of a traditional notion of the Republic of Science and sends a strong message to a wider public that they should stay out of the everyday decision making process of science. I don't think this is the message Nowotny intends to send, but it is a consequence of such a tight definition of a scientific public.

Nonetheless, Nowotny creates a further barrier by constructing a tight definition of a lay person. "By definition," Nowotny (2010, 15) notes, "experts have knowledge and skills that lay persons lack, so that their relationship is characterized by an epistemic...asymmetry. This inequality is structural." However, this relationship need not be stagnate. When Newton published *Principia* Robert Iliffe (2003) suggests very few if anyone understood its meaning and implications and some even doubted if Newton understood it because he did not provide experimental proof and verification of his ideas. Newton's ideas began with a scientific community of one and as Iliffe (2003, 51) notes "Newton's credibility and the authority of the book, was coextensive with his capacity to generate disciples." The success of Newton's theories was in his ability to take "lay people" and help them become experts. It is this process of becoming that Nowotny neglects. She continues to limit the role of lay people when she says those non-specialists who disagree with the processes and outcomes of science they can "opt for protest or they may even seek another, more constructive form of loyalty, by voicing their criticism within the science system and engaging experts in open dialogue. They can also...switch to the political system that offers them voice, the recognized possibility to articulate their dissatisfactions" (Nowotny 2010, 15). Another option is for lay people to become experts and force the system to change in its priorities and decision making process. This is not an unprecedented alternative. AIDs advocates in the 1980s and 1990s did just this. Unsatisfied with traditional testing protocol using a placebo, advocates forced the protocol to be changed in order to make sure all AIDs victims received new medication and not just a placebo. What I am advocating here to prevent the further erosion of the coupling of science with democracy is to open up the gates of the Republic of Science and make it less a Burkean, Rousseauian or Lockean Republic and more like a radical democracy in which more people are involved in the making of science. I am more of an advocate of Dominique Pestre's (2010, 48) vision when he writes "do not limit the process to scientists and engineers; have lay people be full members from the beginning; and have them contribute and co-produce expert knowledge directly." Moving from layperson to expert is a complicated matter and it is the issue that will further strengthen the coupling between science and democracy. It is for this reason that I now turn to the issue of expertise.

5.3 Becoming an Expert

In *The Politics of Nature: How to Bring the Sciences into Democracy* Bruno Latour (2004) shares with us the origins of expertise. Like so many dimensions of Western culture they begin with Plato's allegory of the cave. Plato's allegory "allows a constitution that organizes public life into two houses. The first is the obscure room... in which ignorant people find themselves in chains, ...communicating only via fictions projected on a sort of movie screen; the second is located outside, in a world made up not of humans but nonhumans, indifferent to our quarrels, our ignorances, and the limits of our representations and fictions. The genius of the model stems from the role played by a very small number of persons" (Latour 2004, 13–14). Of course Plato is one of these small numbers of persons as are other philosophers. Latour does not deny that philosophers would still remain part of the elite group, but certainly scientists would be included today. This small group has the power to "move back and forth between the houses. The small number of handpicked experts...can make the mute world speak, tell the truth without being challenged, put an end to the interminable arguments through an incontestable form of authority that would stem from things themselves" (Latour 2004, 14). This traditional notion of expertise is very stagnate and passive. All, including nonhumans, who are not part of the elite travelers between the two houses must submit to the opinions (truths), perspectives (laws), and ways (methods) of the elite. Without these mediators the two houses of human society and nature could never coexist. There would be anarchy and certainly none of the progress the human house made in taming nature. The masses would still be in the cave plotting ways to assassinate the philosopher/kings who know the meaning and importance of the light.

Latour, naturally, rejects the divided houses, rejects the notion that there is such a house as "nature" and seeks to find ways to change the relationships between humans (expert and not) and nonhumans (laboratory equipment, animals, machines, vegetation, inanimate objects). In place of the traditional kingdom of two houses, Latour wants to, as do I, create a revolution in which the monarchy of Plato's allegory is replaced by a parliament of things. This parliament of things is not stable and changes frequently depending on what the issue, problem, or concern is. Instead of two houses making up of a kingdom there should be two houses making up a parliament. The lower house of this parliament will take into account and the upper house will concern itself with putting into order. The lower house takes into account all possible voices involved in a problem. No voice is denied access including those of animals, inanimate objects, laboratory inventions, techniques, and vegetation. These voices are never denied because they may provide a solution to a problem. To adopt a traditional notion of science only the scientist is allowed to speak, his (and traditionally it is a man) opinion is the voice of nature since he is the modest witness who orders the world and speaks for all. In a parliament of things the modest witness is one, weak, voice among many. Once all these voices are gathered they begin "crowding up against the gates" of the upper house where the ordering of the problem begins (Latour 2004, 165). It is the upper house who decides who amongst the crowd will be admitted, tested, probed, questioned, organized in order to construct a theory eventually giving meaning to a problem. When members of the crowd are selected the rest do not just go away, never to reappear again in life. If the initial solution to a problem fails then the gates of the upper house must be opened again, and even if the first solution proves a right one other problems will persist and these problems will not require the same upper house members nor the same members from the lower house crowd. Once the crowd members are selected then the real work begins. "It is not an easy task to transform the inarticulate mutterings of a multitude of entities that do not necessarily want to make themselves understood" (Latour 2004, 168). After all, how does one get an animal, laboratory instrument, plant, or inanimate object to speak to a human if they hold a solution to a problem? The key is the harder one works in the upper house the more likely something will emerge. Work of the upper house is "a blend of skills: an ingenious innovation is developed by clever engineers, one class of beings is substituted for another by bold scientists… accommodations are made behind closed doors, simulations are produced by means of calculations, cold diplomacy is accompanied by the occasional moment of enthusiasm to warm up this improbable heap of compromises" (Latour 2004, 176). Truth is not the outcome of this process of diplomacy in the upper house and negotiations with the silent. A theory, a representation, an ordering explaining a path through or around a problem is presented and eventually adopted or if need be rejected and the whole process of selecting members of the upper house begins again. What emerges from Latour's vision of how science is conducted is not very amenable to neoliberal thought because as Wendy Brown (2015, 44) notes, "Neoliberalism is the rationality through which capitalism finally swallows humanity." There is no room for a parliament of things in the Neoliberal Thought Collective, the Platonic allegory tradition suits them just fine since they handpicked themselves to be the experts, the voices of truth and power. Those interested in creating a world that stretches beyond neoliberal tyranny will find Latour's parliament of things helpful. The work, though has only begun. How does one not only become a member of the lower house, a voice to be heard, but one who is able to get through the gates of the upper house in order to shape science in the laboratory and science policy? Submitting to the "smugness of ignorance" that neoliberal thought promotes is certainly not an answer. This only guarantees one will be a passive conduit still relegated to the cave. How does one become an expert among an array of experts of things? This is where Harry Collins' and Donald Evans' work comes into play.

Collins and Evans offer what they refer to as a third wave of science studies in which the defining Studies of Expertise and Experience (SEE) is developed. The first wave of science studies would consist of sociologists of science such as Merton and Ben-David who constructed expertise in a traditional manner outlined by Latour above. In the first wave of science studies "a good scientific training was seen to put a person in a position to speak with authority and decisiveness in their own field and often in other fields too" (Collins and Evans 2002, 239). Today this is a privilege medical doctors and economists often claim. The second, which is in no way invalidated by the third wave, began to challenge the first wave and discuss ways to challenge the notion that scientists have special access to this thing called nature or truth. Second wave sociologists of science including Collins questioned this authority granted scientists and began to plant scientific authority firmly in historical, cultural and philosophical ground that was shaped and formed by matters of social class, gender, race, ethnicity, politics, religion, sexual orientation, and institutional settings. As new members of "the third wave of science studies" Collins and Evans wish to know "'if it is no longer clear that scientists and technologists have special access to the truth, why should their advice be specially valued?' This, we think, is the pressing intellectual problem of the age." And so do I because it ultimately involves the educating of people to seek expertise and become a part of an expert group without necessarily becoming a member of that group. It is also the pressing problem of the age because it serves as a way to challenge the tyranny of neoliberal thought and the dominance of economists in policy matters in government and education at all levels.

Collins and Evans introduce a model in which there are a core-set of experts, a core-group or scientific community, and the public or lay people. It is the core-set where the controversies usually are played out and the tension is thickest as notions of truth are debated, interpretations of data contested, and consequences of meaning are constructed. These tensions often remain within this core-set throughout the development of a theory or solution to a problem. These tensions are specific in nature pertaining specifically to the construction of a theory or definition of a problem that they rarely are removed from the core-set level and make it to the core-group level let alone the level of the public. "Core-scientists," Collins and Evans (2002, 246) suggest, "are continually exposed, in the case of dispute, to the counterarguments of their fellows and, as a result, are slow to reach complete certainty about any conclusion." It is this tension however that makes this core-scientists' work vibrant and meaningful. It is what generally makes science a productive field of knowledge. However, this is often misunderstood by the public and is an aspect laypeople need to understand if there is to be movement from the public to core-sets. Collins provides an example of this tension in the core-set that was misconstrued in the realm the public. During the email scandal dealing with climate change. Climate scientists emails revealed that one scientist believed that "Our observing system is inadequate" and an attempt to "hide the decline" of earth's temperatures which was described as a trick. On the surface this sounds like serious admissions of an inadequate way to test environmental changes and an attempt to manipulate data to support a specific research and political agenda. This is how political opportunists and human climate change deniers took it, but Collins points out some very important core-set details that reshape these comments in a new light. Collins (2014, 90) explains that in regards to the first statement it was in reference to "our inability to effectively monitor the energy flows associated with short-term climate variability" and in regards to the "'trick' that was mentioned was scientists' normal way of talking about a 'neat trick' for accomplishing some technical transformation, while the 'decline' was in tree rings and not temperature." Any attempt to become experts in science will require lay people to understand the nuance and specifics of the language used to explain the debate involved within the cultures of core-sets.

Core-groups are the broader scientific community. This can include scientists who are part of a specific field of knowledge but not involved in a specific debate or research agenda. It can include scientists from one scientific field when the debate or agenda is in another field of knowledge. It can also include sociologists, philosophers, historians, and anthropologists of science. "A core-group is much more solidaristic group of scientists which emerges after a controversy has been settled for practical purposes." This group looks at the outcomes of the core-set science and sees the emergence of a consensus and may even know a lot about the science involved in the debate and controversy, but they do not see, because they are not involved in the actual research process, the continued debates and controversies Collins and Evans mentioned regarding the core-set group. Often the core-group becomes the spokespersons for the core-set simply because they are regarded as scientists and knowledgeable of the specific scientific issue. This is true for the current debate between evolutionary biologists and creationists. In the public debate evolutionary biologists are rarely in the "limelight" but instead a core-group represents them including people like Neil deGrasse Tyson and Bill Nye. In both cases they are not part of the core-set of scientists. Even as an Astrophysicists, if he is not involved in specific research that is connected to evolution, deGrasse Tyson is not a core-set scientist but a core-group spokesperson. Bill Nye is not even considered a scientist but a person with a mechanical engineering degree and a television background, but he has become the most noted advocate for evolutionary biology.

Then there is the lay people. This group is simply those people who are not involved in the core-set research nor are they scientists. However, it does not mean they are not experts. In fact, when science is played out in the political arena they may be more expert than the scientists. Collins and Evans introduce the sheep farmers of Cumbrian whose livestock were suffering from fallout from the Chernobyl meltdown. The farmers knew exactly what was happening to their sheep, to the soil, the cause of the problem, and what course of action should be taken. "The scientists, however, were reluctant to take any advice from the farmers" (Collins and Evans 2002, 255). The farmers did not speak the scientists' language and they were not viewed as experts. Compare this to the AIDS activists and Collins and Evans (2002, 256) write that the activists "had to learn the language of science if they were to represent the interests of the wider community within the clinical trials process." They did and because of it changed the drug trial process.

Given this arrangement between core-sets, core-groups, and lay people, who should be involved in the decision making process about scientific research? According to Collins and Evans scientists should not be given any special rights in the decision making process simply because they are scientists. This is a major break from wave one of science studies. This does not mean that scientists merely have opinions and their voices should be ignored. It means that who should be involved in the decision making process should not be closed off to exclude non-scientists and scientists should not be included simply because they are scientists. It also means that lay people should be included simply because they are citizens affected by environmental and medical problems. For Collins and Evans (2002, 254) there are three types of expertise that non-scientists possess. There is no expertise or the lay person who cannot contribute anything to the discussion of any scientific topic. Collins (2014, 116) in his own work will refer to this as ubiquitous expertise in which a lay person can "read the literature" and believe that this gives them "a feeling that we are acquiring deep knowledge, but it is an illusion because we have nothing to tell us what parts of the literature to take seriously and what to ignore, to know this one needs interactional expertise." Interactional expertise means possessing enough knowledge within a field to "interact interestingly with participants and carry out a sociological analysis" (Collins and Evans 2002, 254). Finally there is contributory expertise or the ability to "contribute to the science of the field being analysed" (Collins and Evans 2002, 254). There are plenty of interactional experts in the humanities and social sciences including Collins. There is Katherine Hayles whose work with computer scientists and neuroscientists is helping usher in a new era of digital humanities. Then there are cases of contributory experts such as Thomas Kuhn, Peter Galison, and Donna Haraway who have contributed to scientific fields of knowledge before contributing to science studies fields. The issue for curriculum studies is how do we move more lay people including ourselves from the status of no expertise or ubiquitous experts to becoming interactional experts? This is one of our greatest challenges as we face the leviathan called neoliberalism. Neoclassical economists who support a neoliberal agenda already have their interactional expertise and like the scientists of the first wave of science studies they feel a natural right to speak authoritatively about everything. This explains in part why neoliberal thought has made unfortunate inroads into the fields of education and in the everyday decision making of universities. Neoliberals such as Gary Becker brazenly declare they have created a theory of everything and very few people except maybe a few neoclassical economists who do not adhere to a neoliberal ideology challenge them. Instead, the layperson accepts what the economist says as if it were Truth and inevitable with no alternatives to challenge such authoritative and authoritarian declarations.

There are a few examples to draw from within the field. I would consider a few curriculum studies scholars to be already there waiting for more of us. There is Delese Wear in the medical humanities, David Blades and Jayne Fleener in the sciences perhaps, Peter Appelbaum, Liz de Freitas, William Doll, and Brent Davis in Mathematics, and the Goughs, Annette and Noel in Environmental Science. Established scholars no doubt, but not enough to push back neoliberal thought and reunite science with democracy. Besides making our own efforts to become more interactional experts in the sciences, we should begin to make a more concerted effort to encourage teachers to become not only interactional experts in the fields of curriculum studies and pedagogy but also in the sciences. There is no reason humanities teachers including literature and history teachers cannot become interactional experts in the sciences. Interdisciplinary work connecting literature to the sciences is desperately lacking in public schools and can become one way to attack the neoliberal attempt to bury schools in ignorance. History is a natural site to teach about science since it is a topic strangely abandoned by public school history curricula. This is the case in part because the sciences devalue the history of their fields, but the history of science is another way to create interactional experts. The best way to create interactional experts is to find ways to get teachers into the laboratories and into debates and controversies dealing with specific scientific issues. Eventually this will help teachers get students involved as well. This alone will help counteract neoliberal attempts to promote ignorance in public schools and undermine efforts to inform and educate an actively involved citizenry that will seek out to become interactional experts in fields of knowledge of their choice. This is the only kind of choice theory I will accept from neoliberals, but this is not the kind of theory they peddle.

How do we move from ubiquitous experts in science to interactional experts? Ludwik Fleck offers a way. Most have, even in the field of science studies have not heard of Ludwik Fleck, but one can argue he marks a beginning of the rise of science studies. Kuhn briefly mentions and acknowledges him in his groundbreaking work, but most in the field ignore him. Fleck is important to insert here and introduce for three reasons. The first reason I just mentioned. Second, he introduces the idea of thought collectives to scientists from the sociology of knowledge. What better way to challenge the Neoliberal Thought Collective than to replace it with an alternative thought collective or at the very least offer up a more democratic way to think about thought collectives and how they work? Third, Fleck is one of the few warning us about the constant threat of decoupling science from democracy. Where Kuhn, Polanyi, and Hayek assumed the connection was natural, because such a viewpoint, fit their agenda of limiting democratic involvement in science, Fleck realized that science and democracy were not natural fits that could be assumed as a given connection. Perhaps, Fleck was aware of the importance of the connection between science and democracy because when he was writing his book in 1935, *Genesis and Development of a Scientific Fact*, he was in Poland and constantly reminded of the threat of Fascism to the direct West and North and Soviet Stalinism to the East.

Fleck believed that every field of knowledge, no matter how sophisticated and esoteric or common and popular it was, there was a thought collective. It was within the thought collective that people interested in joining the organization, club, or field of knowledge learned how to think, write, and interact with fellow members of the group. "At a certain stage of development," Fleck (1935/1979, 107) believed members learned, "the habits and standards of thought" that "will be felt to be the natural and only possible ones." Certainly one of these habits and standards of thought was how to define Truth and see a fact. "Truth," Fleck (1935/1979, 100) suggested, "is not 'relative' and certainly not 'subjective' in the popular sense of the word. It is always, or almost always, completely determined within a thought style." It is this style that constructs Truth and facts as "an event in the history of thought" (Fleck 1935/1979, p. 100). This thought style is a psychological construct created within the collective and passed in and through the individuals and can be defined as "directed perception, with corresponding mental and objective assimilation of what has been so perceived" (Fleck 1935/1975. 99). The directed part refers to the fact that members of the collective are educated as to how to see the world and the corresponding mental and objective assimilation aspect refers to the fact that how collective members are taught to see can be referenced and confirmed in a reality whether that is a reality of how a molecule functions or how knowledge functions. It does not matter what the topic is. All thought collectives and their corresponding styles are historically contingent and it is difficult for anyone who is interested in looking back to earlier thought collectives to understand how it functioned and saw the world. The only way to understand it is to abandon current thought styles and adopt a previous style. Fleck stressed the historical contingency of thought collectives and styles because even contemporary ones will be outdated and at the very least altered and future anthropologists and sociologists of knowledge will be tempted to sneer at our current limits, notions, and assumptions, but they will not be able to understand the culture unless they adopt its style.

Within all thought collectives is a general structure that is very similar to what Collins and Evans developed above. "The general structure of a thought collective," for Fleck (1935/1979, 105), "consists of both a small esoteric circle [Collins and Evans' core-set] and a larger exoteric circle [core-group] each consisting of members belonging to the thought collective." The esoteric members are what Fleck calls the "initiated" or the trend setters who establish the standards, the styles, and cultures of conduct. In science the initiated would be the members of a specialized subfield with a discipline of knowledge who understand the inner-workings of that specific area of knowledge and wish to convey to the other members of the importance of their work and the protocol by which their work should be approached. The exoteric group is a little different from Collins and Evans core-group in one important way. Whereas a core-group in science is usually scientists who understand the nature of a subfield but do not necessarily know the details and controversies defining the research within the subfield. For Fleck the exoteric group can include anyone including the general public who may be interested in a topic or a discipline and it is this slight difference that makes Fleck very important for my discussion. Fleck observes that esoteric groups may be the trendsetters of research protocol and the specialists within the field but they "are more or less dependent, whether consciously or subconsciously, upon 'public opinion', that is, upon the opinion of the exoteric circle. However, the tendency of an esoteric circle will be to close itself off and limit contact with the exoteric circle. The "public" has "no immediate contact with the powerful dictators forming the esoteric circle. Specialized 'creations' reach them only through what might be called the official channels of intracollective communications, depersonalized and thus all the more compulsive" (Fleck 1935/1979, 108). The vitality of the thought collective in terms of its openness is not based on the actions of esoteric circles because their tendency is to be dictatorial in controlling the production and flow of knowledge. It depends on the strength of the exoteric circle. "If the masses occupy a stronger position," Fleck (1935/1979) asserts, a democratic tendency will be impressed upon this relation …. If the elite enjoys the stronger position, it will endeavor to maintain distance and to isolate itself from the crowd. Then selectiveness and dogmatism dominate …. This is the situation of religious thought collectives. The first, or democratic, form must lead to the development of ideas and to progress, the second possibly to conservatism and rigidity." Fleck's sociology of knowledge describes well the current state of the relationship between science and democracy. The Neoliberal Thought Collective has introduced its style to the realm of science and the esoteric circles within science have corporatized themselves, shutting out the exoteric circles as much as possible to limit, if not completely eliminate, their influence on important policy and research decisions. It can, and I submit it should be, an important incentive for curriculum studies scholars along with teachers and students to join scientific exoteric circles of their choice and, in the name of democracy, enter into more of the esoteric circles demanding a more openness that will shed light on the darkness of neoliberal thought and reconnect science with democracy. This is our challenge and should be our goal as educators and curriculum studies scholars.

References

Barron, C. (Ed.). (2003). A strong distinction between humans and non-humans is no longer required for research purposes: A debate between Bruno Latour and Steve Fuller. *History of the Human Sciences, 16*(2), 77–99.

Brown, W. (2015). *Undoing the demos: Neoliberalism's stealth revolution.* Cambridge, MA: Zone Books.

Collins, H. (2014). *Are we all scientific experts now?* Cambridge: Polity (F. Bradley & T. Trenn, Trans.). Chicago: University of Chicago.

Collins, H., & Evans, R. (2002). The third wave of science studies: Studies of expertise and experience. *Social Studies of Science, 32*(2), 235–296.

Derrida, J. (2004). The principle of reason: The university in the eyes of its pupils. In J. Derrida (Ed.), *Eyes of the university: Right to philosophy* (Vol. 2, pp. 129–155). Chicago: The University of Chicago.

Fleck, L. (1935/1975). *Genesis and development of a scientific fact.* Chicago: University Of Chicago Press.

Fleck, L. (1939/1979). *Genesis and development of a scientific fact.*

Hands, D. W. (2001). *Reflection without rules: Economic methodology and contemporary science theory.* Cambridge, UK: Cambridge University Press.

Harrison, R. (2014). *Juvenescence: A cultural history of our age.* Chicago: The University of Chicago.

Hayek, F. (1948/1980). The use of knowledge in society. In F. Hayek (Ed.), *Individualism and economic order* (pp. 77–91). Chicago: The University of Chicago.

Husserl, E. (1954/1970). *The crisis of European sciences and transcendental phenomenology.* Evanston: Northwestern University Press.

Iliffe, R. (2003). Butter for parsnips: Authorship, audience, and the incomprehensibility of the Principia. In M. Biagioli & P. Galison (Eds.), *Scientific authorship: Credit and intellectual property in science* (pp. 33–65). New York: Routledge.

Kant, I. (1979). *The conflict of the faculties.* Lincoln: University of Nebraska Press.

Latour, B. (2004). *Politics of nature: How to bring the sciences into democracy.* Cambridge, MA: Harvard University.

Mirowski, P. (2011). *Science mart: Privatizing American science.* Cambridge, MA: Harvard University Press.

Mirowski, P. (2013). *Never let a serious crisis go to waste: How neoliberalism survived the financial meltdown.* London: Verso.

Nowotny, Helga. (2010), The changing nature of public science, Helga Nowotny, Domnique Pestre, Eberhard Schmidt-Assman, Helmuth Schulze-Fielitz, Hans-Heinrich Trute, The public nature of science under assault: Politics, markets, science, and the law, 1–27. Berlin: Springer.

Pestre, D. (2010). The technosciences between markets, social worries, and the political: How to imagine a better future. In H. Nowotny, D. Pestre, E. Schmidt-Assman, H. Schulze-Fielitz, & H.-H. Trute (Eds.), *The public nature of science under assault: Politics, markets, science and the law* (pp. 29–52). Berlin: Springer.

Pestre, D. (2012). Concluding remarks. Debates in transnational and science studies: a defense and illustration of the virtues of intellectual tolerance. *British Society for the History of Science, 45*(3), 425–442.

Pignarre, P., & Stengers, I. (2011). *Capitalist sorcery: Breaking the spell.* New York: Palgrave.

Plehwe, D. (2009). Introduction. In P. Mirowski & D. Plehwe (Eds.), *The road from Mont Pèlerin: The making of the neoliberal thought collective* (pp. 1–42). Cambridge, MA: Harvard University.

Polanyi, M. (1969/2002). The republic of science: Its political and economic theory. In P. Mirowski, & E.-M. Sent (Eds.), *Science bought and sold: Essays in the economics of science* (pp. 465–485). Chicago: The University of Chicago.

Proctor, R. (2008). A missing term to describe the cultural production of ignorance (and its study). In R. Proctor & L. Schiebinger (Eds.), *Agnotology: The making & unmaking of ignorance* (pp. 1–33). Stanford: Stanford University Press.

Schelling, F. (1802/1966). *On university studies* (E. S. Morgan, Trans.). Athens: Ohio University.

Part III
Interlude Three: Ghosts in the Resistance and Bones in the Soil

Jodi Byrd's (2011, p. xxxii) words haunt me: "For those within American Indian and indigenous studies, postcolonial theory has been especially verboten precisely because the 'post-,' even though its contradictory temporal meanings are often debated, represents a condition of futurity that has not yet been achieved as the United States continues to colonize and occupy indigenous homelands." They haunt me as an heir to European colonialism and a progeny of continued USA imperialism at home and abroad. Everything that is built in the USA and in the name of the Western world is built on top of the bones and ashes of indigenous people. This is why Byrd says that postcolonial theory is forbidden in the circles of Indian and indigenous studies. Postcolonialism too is built on the soil where indigenous people were massacred and eventually erased from memory; constructed as savages needing to be "saved" by the Western ways of Christianity and capitalism. The limits of postcolonial thought have been found and like so much of postmodern thought it makes me wonder how progressive is it if indigenous people are not recognized as a part of the literal foundation of all Western Hemispheric theory and practice? Byrd (2011, p. 229) ends her work with this thought: "it is time to imagine indigenous decolonization as a process that restores life and allows settler, arrivant, and native to apprehend and grieve together the violences of U.S. empire." One starting point to this restoration of indigenous ghosts and massacred bones is to recognize the rise of a postcolonial science that challenges current imperial practices so often buttressed by science matters. As with other matters scientific, this curriculum scholars are yet to do.

Outside of science, two curriculum scholars have already laid the foundation for the creation of dialogue between indigenous and postcolonial scholars which allows me to believe perhaps postcolonial theory will be soon removed from the forbidden list. I am referring to the work of Eve Tuck and Rubén Gaztambide-Fernanádez. For the heirs of settler colonialism, and I would add settler capitalism and its research arm, colonial science, their work is not meant to be comfortable or inviting. In fact, are we invited at all? Who can question their hesitations or doubts? How many times have settler colonialism coopted then claimed as their own something indigenous or colonized people created? How many times after starting a movement, a culture, a thought have indigenous and colonized people been push aside after their work has been claimed by colonialists and the memory of the origins erased? The goal of Tuck and Gaztambide-Fernández is to protect indigenous knowledge from settler colonialists. What role is there for those of us who are heirs of settler colonialism and interested in not continuing the traditions that erased indigenous people from the earth and colonized the rest in the name of a god, capital, and/or science? The immediate response from Tuck and Gaztambide-Fernández in their important article, "Curriculum, Replacement, and Settler Futurity," is we are not invited. Their theory in part is one of refusal. "There isn't an easy ending. We anticipate that even with all of these refusals and exactions, this article is just as likely as any other to be incorporated and absorbed...We wonder who will notice when the Natty Bumppos of the field will both praise and dismiss, remove and replace, take what is necessary and position themselves, once again, as the true 'native'." In this warning to stay away we heirs see our possible role. To not erase, not claim our perspective as native; to accept we are the settlers who destroy not in the name of progress but idolatry, greed, lust, and murder. It can be our role to change this history of death and destruction and resist the empire that erases and condemns to death those who do not fit the description of "native" Americans even if they are Native.

In the midst of Byrd's warning of the strained relations between indigenous and postcolonial scholars, Tuck and Gaztembide-Fernández's work symbolizes a challenge to this tension. In co-authoring this piece they indirectly offer a pathway to relieving this strain as they come together and recognize the need not to erase the indigenous that haunts settler colonialism and bleeds from the soil at every point of reference in the USA. Their coming together offers only a stronger challenge to USA imperialism, a force that long ago needed to become a ghost and never taint the soil again. The intellectual strength of Tuck and Gaztembide-Fernández's work is not how can it be cited, coopted, erased, or challenged, but how will they continue to dialogue and coauthor works that challenge and change settler colonialism. Those of us who are of Natty Bumppos's lineage, our task is to accept we are not invited to this party and if we try to peer into the window to see what is happening at the party then we are the intruders, the uninvited who are illegal. So from a distance Eve and Rubén best wishes on more intellectual exchanges and challenges.

The two chapters, "(Post) Colonial Science" and "Working out way back: Colonial Science in light of Postcolonial Thought," that follow deal with postcolonial science. It is a topic curriculum scholars barely acknowledge as existing, yet I hope to demonstrate it is a field bearing so much fruit. The first chapter I depend on the work of Sandra Harding, Paul Gilroy, and Sylvia Wynter to determine what it is postcolonial science presents to curriculum scholars as challenges and insights into theory. The second chapter is applied postcolonial science. I focus on the work of Bleichmar (botany), Schiebinger (more botany and biology) and Carney (agriculture and engineering) to demonstrate how postcolonial science tells a different story about science and the limits of notions of progress, objectivity, and neutrality. Colonialism and slavery were costly results for people to pay so a myth could be invented that rationalized the power of the West to conquer the world, the inevitability of capitalism as the only viable economic order, and science as a mode of inquiry based in universal truths. As imperialism, capitalism, and Western science garnered the most attention, other paths were developed and different stories were being made. Postcolonial science is just beginning to recover them and tell them anew. In the foreground of these chapters and between the words, rest Byrd's warnings. Is a postcolonial science even possible? This question will and should haunt these chapters because they are, like anything else created in the Americas, built on the remains of first nations' people and a past purged from memory.

References

Byrd, J. (2011). *The transit of empire: Indigenous critique of colonialism.* Minneapolis: University of Minnesota Press.

Tuck, E., & Gaztambide-Fernádez, R. (2013). Curriculum, replacement, and settler futurity. *Journal of Curriculum Theorizing, 29*(1), 72–89.

Chapter 6
(Post)Colonial Science

All the problems that we attempt to solve by means of science and technology—whether 'the use of better methods of birth control'…antiballistic missile system…new foods and better ways of growing them…novel ways of reducing or disposing of waste…steps to control disease…or housing and transportation…return to haunt us—Sylvia Wynter. 1997. "Columbus, The Ocean Blue, and Fables that Stir the Mind" 145

As I begin to write this chapter and the next one on postcolonial thought and science, it is not hard to find evidence to support what Sylvia Wynter was writing about in 1997. In an effort to combat the potential of mosquitos to spread the zika virus the state of South Carolina spread chemicals that not only apparently reduced the mosquito population but also killed millions of bees, a population already decimated by genetically modified foods and other human actions. The television is filled with commercials from law offices asking if you took this drug, fill in the blank and take your pick of a long list of pharmaceuticals, and if you are now experiencing the following medical ailments ranging from kidney failure to lung and heart problems please contact fill in the blank of the name of a law office. The Syrian civil war, now involving Daesh, forces loyal to Bashar al-Assad, forces revolting against the Assad regime, Great Britain, the United States, and Russia continues into its 7th year, began as a struggle over water, a growing concern as a result of climate change and population growth. The role of science in the development of Western empires has been a topic of concern of historians of science for decades. Originally science was seen as a mechanism for justifying the creation of an empire. Science was a symbol of Western superiority and development. Because of science Western monarchies supposedly possessed the best means of cultivating lands, controlling rivers, forests, and plants, educating peoples, and dictating policies. Science became the rationale for eliminating indigenous cultures and justifying the conquering of 84% of all the earth by 1930 when Western imperialism was at its height. The colonial mentality of superiority did not end with the rise of independence and postcolonial movements. There should be little doubt that modernist notions of development and Western superiority still exist. These mentalities are just more complicated it appears.

© Springer International Publishing AG, part of Springer Nature 2018   127
J. A. Weaver, *Science, Democracy, and Curriculum Studies*, Critical Studies
of Education 8, https://doi.org/10.1007/978-3-319-93840-0_6

One example of this complexity is the relationship between science and the post-colonial state of India. The historian of science Itty Abraham demonstrates that after independence the idea of postcolonial science in India had two distinct meanings. The first was a view held by Mahatma Gandhi and his supporters who believed that in order to create an independent India traditional Indian approaches to life had to reemerge and often science was viewed as an impediment to the resurfacing of these traditions. The second, and most prominent, approach to emerge was one that held that science was a means to become an independent nation. Science, with a concept of basic research, rationality and objectivity serving as its anchor principles, would help India emerge as a developed and modern country. This gave rise to the development of India's own science traditions in astronomy and biology as well the eventual creation of their own atomic weapons. The belief at the time, as it remains today, was that if a nation were to be taken serious as a developed nation it would have to create, among other things, a basic research agenda that could be verified as legitimate by the international (mostly Western) scientific community and contribute to the body of knowledge that creates economic opportunities for growth not only for its own nation but the world in general. Traditionally, a nation such as India during the height of colonialism was seen as the site where basic research was applied thereby continuing a binary process of viewing basic research as superior to applied science. As a result, Abraham (2000, 54) suggests this binary of separating basic and applied science "helps us trace the origins of the presumed inferiority of the colony in the hierarchy of knowledge production even as it points to the long history of modern science as a transnational practice." More importantly, this binary encouraged the new nation-state of India to create a set of goals that placed international colonial acceptance over the everyday needs of the people of India. For example, the leaders of independent India viewed the making of an atomic bomb as important because it was a symbol of basic research and when successfully detonated the bomb would serve as a public relations declaration that India is now a developed and modern nation. Meanwhile the needs of the people were ignored. The only difference then between colonial and postcolonial science in this example becomes this: science in the colonial era was done in the name of an empire and in the postcolonial era it is done in the name of the nation-state.

Since the rise of the notion that basic science and the philosophy behind science (rationalism and objectivity) were the reasons for Western superiority and the means by which nations developed their economies and societies, historians of science are beginning to challenge these assumptions while creating alternative ways of viewing science and the history of colonialism. Part of the original thesis of the role of science in colonial empires was the belief, which justified and rationalized the concept of objectivity, that science was not a driving force towards the formation of colonies rather religion was. Science was merely an innocent bystander in a system driven by the evangelical desire to convert the souls of Africans and indigenous peoples of the new world to the word of Christ. If these peoples were to heed the call to Christ they would be saved and classified as human and if they did not then they were justly given their warning and enslavement and/or military conquest were legitimate outcomes for the heathens, savages, and infrahumans. Harold Cook is one historian who has recently challenged the innocence of science. In his work, *Matters of Exchange: Commerce, Medicine, and Science in the Dutch Golden Age*, Cook (2007, 4) believes any attempt to suggest the driving force to create colonial empires was religion "is inadequate." It was commerce along with science that drove imperial desires. Roy Macleod (2000, 2) writing before Cook presents a similar line of thinking when he stated "Science, accompanied transoceanic colonization, and in shaping the economic destiny of the West, defined the relationship between the 'old' and the 'new.'" This new relationship not only created imperial policies and individual nation-state drives to explore. It created a new way of thinking, seeing, and interacting (with)in the world. It created the need for "a knowledge appertaining to a detailed acquaintance with objects" such as spices, species, and other objects of commerce and learning (Cook 2007, 17). There grew a need to know what could be "useful" in the "new" world for the "old" world. There is no coincidence that with the rise of capitalism also grew the rise of botany, biology, astronomy, and chemistry as well as zoology; all applied sciences with tremendous economic profit.

Suman Seth further challenges the notion of science as innocent server of the world and mechanism for modern development. Seth (2009, 373) suggests the "idea that science and technology were among the gifts that Western imperial powers brought to their colonies was an integral part of the discourse of the 'civilizing mission,' one vaunted by both proponents and critics of the methods of colonialism." Religion here does not become the reason for exploration but the rationale to justify the exploitation of what Western eyes saw in the "new" worlds of Africa, India, and the Americas. At the same time science does not become the savior of the "new" worlds from superstition, savagery, poverty, and exploitation. It is the accomplice. "The history of almost all modern science," Seth (2009, 374) suggests, "must be understood as 'science in a colonial context.'"

As a result of this new trend in historiography to construct science as an economic engine for imperial expansion and as a rationale to justify colonial exploitation of non-European peoples, scholars recently have coopted science for the interests and needs of indigenous peoples thereby beginning a process of releasing the grips of colonialism, Western exceptionalism, modernity, and economic exploitation and delinking science from Western biases and assumption as to what should be valued as basic or applied, objective or subjective, neutral or political. This change emerges from what can be labelled numerous ways including a postcolonial, postmodern, postnational, and postcapitalism approach to science. David Chambers and Richard Gillespie offer one way that can be associated with this new approach. They wish to offer a perspective that moves beyond the modern tradition of viewing science as part of nation-state development or colonial exploitation. Instead they seek to focus on the locality of doing science. "To define a science locality" Chambers and Gillespie (2000, 228) posit "is simply to nominate a local frame of reference within which we may usefully examine the role of knowledge and inculcation." This could be local indigenous people affected by a scientific project such as an oil pipeline, an endangered species, or land mass such as a coastal area. The list of possible forms of locality would be context specific depending on the proposed scientific work, the geographical area involved, and the environment impacted. The point for Chambers and Gillespie (2000, 235) is to focus on locality because, and this is part of dismantling the myth of science as neutral and delinking it from a colonial mentality, "all knowledge systems are 'situated' in power relationships, value assumptions, and historical frameworks. As a culturally specific knowledge system…Western science…cannot be accorded a privileged status over indigenous knowledge." This approach of course does not mean somehow Western science is now dismissed as an illegitimate form of knowledge, and to discourage any of the hysterics of those prone to overreact to this new paradigm of science, nor does it mean that we are entering a new era of anti-intellectualism and anti-science. What this new paradigm is advocating is something Western science should have done centuries ago. It simply should have taken into account the needs, knowledges, assumptions, and beliefs of all those people impacted by science. As Timothy Choy (In Seth 2009, p. 379) writes "'science does not leave the building; it simply has to demonstrate that it has particular reasons to stay.'"

In this chapter I will look at the main postcolonial theorists who have argued for the idea of locality: Sandra Harding, Paul Gilroy, and Sylvia Wynter. Each one of these postcolonial thinkers provides insights into how science can be rethought within the context of indigenous peoples and environments and their knowledge and assumptions about their world. These three scholars provide important ways curriculum scholars can interact more in the role of science in society and the work that scientists do throughout the world. Their work becomes more ways in which curriculum scholars can enter into scientific discourses and make an impact on how scientific research is done in a democracy. Often the work of Harding and Gilroy is ignored in curriculum studies and in the case of Wynter her important insights on rethinking science in relationship to the humanities and indigenous peoples are often not covered when discussing her work.

6.1   Saundra Harding and Strong Objectivity: Science from Below

Of the three scholars discussed in this chapter Sandra Harding writes the most about the connections between postcolonial thought and science. Her early scholarship as a philosopher of science focused on women's issues and feminism, but with such works as her edited book *The 'Racial' Economy of Science* (1993) and her own book *Is Science Multicultural: Postcolonialisms, Feminisms, and Epistemologies* (1998) Harding's work made early contributions to postcolonial thought and science studies. Her major contribution to the field of postcolonial science studies is strong objectivity. Succinctly, strong objectivity is the idea that if science is to be done ethically and correctly all the voices, human and non-human, animal and non-animals, have to be taken into consideration before any research can be conducted. If all the voices are not considered in understanding the impact particular scientific research will have, then it cannot claim to be a strong form of objectivity but at best a weak form of objectivity and perhaps even just a form of ideological scientism favoring powerful voices in the world. Embedded in the concept is not only a critique of traditional modern notions of objectivity but also patriarchy, colonialism, Western ethnocentrism, Western exceptionalism, and modern notions of science. In one of her latest books *Science from Below: Feminisms, Postcolonialities, and Modernities* (2008), Harding offers insight into the impact modern science assumptions have on indigenous peoples and the environments of the world. "As long as the modern conceptual framework of philosophy, sciences, science studies, and other research disciplines in the West," Harding (2008, 233) suggests, "leads us to believe that what happens in households around the globe constitutes obstacles to the advance of objective and reliable knowledge" science will never embody its assumptions of serving as a measure of and for progress nor as a mechanism to improve the lives of individuals. In order to create a form of science that works for the needs of all of the peoples, non-human animals and ecologies in the world the major principles of science will have to be rethought. This will have to include a rethinking of the principles undergirding modernity including Western Exceptionalism, triumphalism, progress, binary thinking (Western science knowledge vs. traditional "primitive" superstitution), and the idea of abstract, universal, and neutral objectivity. In this section I will cover each of these points and provide two pertinent examples of how science is being done differently taking up Harding's principles rather than continuing the traditional, modern approach to science.

Western exceptionalism according to Harding (2008, 3–4) is "the belief that Western sciences alone among all human knowledge systems are capable of grasping reality in its own terms…only modern Western sciences have demonstrated that they have the resources to escape the universal human tendency to project onto nature cultural assumptions, fears, and desires." This notion of exceptionalism is not only embedded in modern science but also in how many Westerners view their human role in the environments on the earth as they delegate to themselves the ultimate power to determine what is good for environments and what is not. It should not be a surprise that the questioning of the assumptions of modern science emerges at the same time Anthropocentrism is questioned. This multipronged questioning of similar dimensions of modernity can also explain the United States Christian anti-scientific stance against evolutionary biology and climate change. Christian fundamentalists in the United States have conflated the questioning of their "god" given Biblical right to serve as stewards over god's land as lords over their dominion with postcolonial, feminist critiques of Anthropocentrism. As we will see with the case of climate change matters are much more complicated than this fundamentalist protestant theology would like to have it. There are major differences between this pre-modernist theology and the modern scientific assumption that believes scientific methods are not only the best but the only way to capture reality, although they both proclaim a knowledge monopoly. It is important to note that just because Harding takes to task the modernist assumption that the Western sciences are the only way of knowing, it does not mean she is somehow intellectual allies with the premodern fundamentalist Protestants or a postmodernist. She is attempting to construct an alternative path to these two alternatives to modern science.

Triumphalism is the sister concept of exceptionalism. Embedded in triumphalism is an inerrant and continued arrow of progress that assumes science is always moving towards improvement and any connections between world environmental or medical crises cannot be associated with science, other than science providing the answer to any crisis. "From this perspective," Harding (2008, 4) notes, "Hiroshima, environmental destruction, the alienation of labor, escalating global militarism, the increasing gap between the 'haves' and the 'have nots,' gender, race, and class inequalities—these and other undesirable social phenomenon are all entirely consequences of social and political projects." Triumphalism would be a form of "cherry picking" data. If something were to emerge as positive and successful in a scientific agenda then triumphalism would proclaim it is an exemplar of what science can do in the world, but if something were to go wrong with a scientific agenda then triumphalism would label it as an example of politics or cultural influences impacting the results of science. Triumphalism is simultaneously an extreme form of optimistic thinking and at the same time tragically, irresponsibly, and dangerously naïve.

Both exceptionalism and triumphalism in modern Western science has created a limited binary approach to reality and the possibilities of life. If Western science is the symbol of empirical fact and knowledge then anything that is not under this umbrella must be superstition, magic, or gossip. It should be no surprise that since the emergence of modern Western science in the 1400s over the centuries women such as midwives were accused of witchcraft because their alternative medical traditions were not traditional modern patriarchal forms of knowing, indigenous Americans were called savages and Africans were called infrahumans. These three groups offered different ways of knowing that could not fit the power structures and epistemological assumptions of modern Western science. In regards to women their knowledge became a feminine form of knowing while the masculine knowledge became reliable, objective, and factual. For indigenous people their knowledge became superstitious magic and for Africans theirs witch doctor voodoo. From this binary thinking emerged an erasure of certain voices from the official discourses over knowledge and the sciences. In the case of women, Harding (2008, 199) notes in summarizing the work of Joan Kelly-Gadol that they "have lost status and prestige because of the achievements claimed to make the event or era progressive…. Women appear as missing in action in projects aimed to bring social progress, although careful observation reveals that the action would be impossible without their loss of status and prestige." In other words, the very reason something was labeled as scientific progress was because alternative ways of knowing often dominated by women or other groups were labeled as unimportant, not scientific, or completely erased. In the next chapter when I discuss the history of African rice in the United States, Judith Carney demonstrates that this is exactly what happened to Africans and their vast knowledge of rice. Their knowledge of how to grow rice was superior to the Europeans but as soon as this process was learned by the Europeans the slave plantation system became the most efficient and scientific way to grow any crop in the Western Hemisphere. Another example comes from Londa Schiebinger and her work on colony Botany. When European empires emerged an important part of the economic policy and scientific agenda for these monarchies was to explore, "discover" and extract as many profitable and useful plants from the new lands as possible. Overwhelmingly the explorers were males who often ignored those plants that might have some benefit or use for women. Abortifacient plants were one such example. One of the reasons for ignoring these plants is because most European women could already use the Savin plant if they wished to abort a fetus. A second important reason for ignoring abortifacients was abortion in the early colonial era was a matter for women and their midwives, rarely did male medical doctors involve themselves with the obstetrics and gynecological health concerns including the delivering of children. When eventually the power of midwives was reduced and medical doctors became superior to them, then abortifacients became a topic of debate but mainly to outlaw them and monitor the safety of the fetus in the mother's womb. When explorers came across what was referred to as the peacock plant in Africa they saw very little utility for the plant. This does not mean they did not know what it was used for. As Schiebinger (2004, 228) notes this knowledge "was not cultivated...and not embraced." As a result of ignoring this knowledge, Schiebinger believes the lives of women were threatened because "women depended on this knowledge." In fact when African women were forced into slavery and sent to the Western colonies they often smuggled rice in their hair and the peacock flower on their bodies and used the flower later on when they were raped and impregnated by a plantation owner. Although Western Botanists did not see the value of these plants, African women knew of its importance. As Harding (2008, 200) recognized, "it is often...women who develop, and preserve local pharmacologically useful plants and the knowledge of how to apply them as well as more general health practices and medical treatments." The peacock flower became a form of local medical knowledge used as a mode of resistance even though it was classified and labeled by Western science as non-knowledge and part of African voodoo.

When the underlining assumptions of modern Western science such as exceptionalism, triumphalism, and binary constructs are questioned, strong objectivity is able to emerge and become an important part of how science is done. Harding (2008, 225) raises an important question towards the end of her book: "What would we learn if we started thinking about it [science and technology] and its effects not from the dominant conceptual frameworks, but rather from the daily lives of those groups forced to live in the shadows of such specters—namely those who have benefited least from the advance of modernity's so-called social progress?" What would it take for scientists to do a science from below? Again as it is with other topics in this book it is important for scientists to not overact to these types of questions or to the types of inquiries emerging from science studies. The goal is not to eliminate science from our epistemological field of choices to construct knowledge, rather the goal is to be broader in defining what knowledge is and what science can mean. One way Harding (2008, 41) advocates for a broader notion of science is to encourage everyone involved in science work and policy to "listen carefully to each other in order to understand ourselves and our worlds." Listening would require scientists to understand the indigenous cultures their work will impact and in order to understand how scientific work may impact a culture or environment it will be important to understand how those people in a different culture view the earth, the land, and themselves in relationship to the earth and land. Earlier in this chapter I introduced the topic of the making of the Indian atomic bomb. This weapon became a symbol of India's rise to modern development even though science was not being used to meet the needs of the people, especially those who were living in poverty. Itty Abraham continues his story of the Indian bomb by quoting one of the researchers involved in the project. Once the bomb was completed the scientists had to find a spot to detonate it underground. Eventually they decided on the Thar Desert in western India. One researcher described the Desert this way: "An unexpected bonus as we readied the site for the experiment was our discovery of the great beauty of the desert and its denizens. The Thar bubbled with life…The Bishnois [a nomadic group] for many centuries have acted as the guardians of the animals of the desert, in keeping with their religious beliefs…I do not recall anybody falling ill during that period." (Abraham 2000, 63) Later in this researcher's account, he notes how the team of scientists were not sure exactly where to detonate the bomb in the desert so they sought out the elders of a local village to find a good spot and an elder revealed an abandoned well that could be used. Once the bomb was detonated he noted "'right in front of us, the whole earth rose up as though Lord Hanuman had lifted it.'" (Abraham 2000, 64) Without doubt it was a violent disruption of the local environment, however as Abraham noted that from this narrative modern development and scientific progress the local peoples and animals of this dessert were erased after the pastoral idyllic landscape was described. Abraham (2000, 64) asks "Where did these villagers come from? Where did they acquire their 'local knowledge'? What happened to them after the test? The official narrative remains silent on these questions." We do know that the scientists according to Abraham returned to New Delhi successful heroes. The local sentient population was not a concern to these scientists and the nation-state before the project was planned nor after when modern science was hailed as a successful pathway for Indian economic development. The only time the local population was consulted was when a specific site was needed to detonate the bomb. This example is not what Harding is referring to when she calls for the need to listen carefully to understand others and the impact scientific work will have on everyone. This is an example of modern scientific hubris that proclaims to know all and to know best. For Abraham (2000, 67) the Indian atomic bomb is an example in which "science will only be known by an act of force, not understanding."

Harding suggests other dimensions of what it means to do science from below and develop a principle of strong objectivity. Another aspect of science from below is to accept a particular mantra and work from there: start from the household. Start from the household in defining research agendas, start there to define the meaning of science, start there in order to incorporate the needs of women and children. "Recollect the feminist standpoint mantra to 'start off research and politics from women's lives,'" Harding (2008, 225) urges us, "rather than from the conceptual frameworks of the research disciplines, to create the kinds of knowledge women need and want to empower themselves and their dependents." This though is only the beginning. Science should "start off from the lives of the oppressed, but…not end there, as, for example, do conventional ethnographies…. [A standpoint perspective would] 'study up,' to identify and explain the material and conceptual practices of power which are often undetectable by those who engage in them" (Harding 2008, 225). The reason to start with the household and with the concerns of women is to recognize that "households are the site of one third to one half of human economic activity, including households in modern societies" (Harding 2008, 229). This part of Harding's and standpoint feminists' mantra does run the risk of reducing science to economic utility but it would dramatically alter how we think about the role of science in society. The other part of this mantra, study up, is just as dramatic but not a reduction of science to economic utility. Studying up is a call to investigate what science does, how it is done, and what the power and cultural dimensions of science are. The more non-scientists study up the more of the black box of science will be revealed and no doubt exposed as dangerous. Inevitably the more one studies up, the more dangerous it will be because those with the power to name knowledge, define research agendas, and control the purse strings will not appreciate the intrusive and the inquisitive probes. However, in a democracy no one has the right to limit access to the making of science.

An example, I think, of studying up is found in the work of Myanna Lahsen. Lahsen is an anthropologist who studies climate scientists and their challengers. By studying up and exploring the upper echelons of science and power, Lahsen not only demonstrates the risks of using models to make projections dealing with climate change, but she also exposes the money trail that underwrite the climate deniers. She also rejects binary thinking that reduces this debate over human made climate change to either one believes in climate science or one is a denier. In her research she notes there are scientific skeptics of human made climate change who ask very important questions of the science but do not dismiss the possibility of the existence of the Anthropocene. What emerges from her research is a very nuanced and important contribution to the importance of understanding the process of creating solid research agendas that have impacts on human lives, environments, cultures, and scientific knowledge. By studying up Lahsen avoids the modernist notion of the Aristotelian mean in which two diametrically opposed views are constructed and a false middle is developed to create an alternative that limits possibilities to only three thereby eliminating other possible routes to move beyond the binary of either/ or thinking. What Lahsen, for instance, found in her work on the use of computer modeling is the scientists who create these models sometimes confuse the data they generate from their model with fact. In interviewing these scientists she describes the years of effort teams of scientists put into creating these models and as a result these scientists are more likely to be sensitive to criticisms of their models and more likely to defend the "authority of their models" to name, discover, and define facts (Lahsen 2005, 905). As a result Lahsen (2005, 906) concludes "my participant observation and interview data suggest that model producers are not always inclined, nor perhaps able, to recognize uncertainties and limitations with their own models." Skeptics, a hallmark of science, are necessary in the process of establishing scientific facts in climate science. Not skeptics who deny the existence of climate change but skeptics who ask tough questions of the model makers and the scientists who use the data to draw conclusions about computer projections about future climate changes and their effects on the earth. What emerges from Lahsen's work is a robust and contested contact zone where locality is important, science is debated, challenged, and fought for and facts are not easily accepted but struggled over in order to define not only the present state of science but the future course of policy development and the health of the planet.

Over the 6 years, Lahsen studied the climate scientists she not only recognized that there were skeptics of this work who offer up important insight on how to make the models better and how to draw important conclusions from the data in order to form facts, She also interviewed skeptics often referred to as "the trio". From these interviews one can see the different cultures in the scientific communities, the role of politics and economics in climate change deniers, the importance of paradigm shifts in how different generations do science, and the very important shift Western science has made moving from a physics dominated worldview to one defined more by the biological sciences. "The trio", Frederick Seitz, Robert Jastrow and William Nierenberg, all worked at the Marshall Institute, a conservative funded center advocating free market solutions to world problems and deniers not of science but of climate science in general and many of the conclusions climate scientists draw from their data especially the idea that climate change is an immediate concern and the world needs to act now (Lahsen 2008, 211). They are all physicists and as physicists they are not trained climate scientists but as Lahsen points out they probably developed an interest in challenging climate science for at least two reasons, opportunities for physicists are dwindling and the culture of physicists since World War Two has been they are kings of the knowledge hierarchy and as a result they can study and comment with authority on any scientific topic including climate science. "At a deeper level," Lahsen (2008, 214) suggests "then, it is less that the trio is against having practical and political priorities shape science agendas than they do not agree with the new priorities shaping present science funding, including the demotion of physics." Some might argue that Lahsen is acting like ethnographers often do and not going as deep as studying up requires. I would submit though in her ethnographic research Lahsen is studying up at the most local of levels where a science from below should always start. She begins her work with a concern for the earth and the potential devastation climate change can have on every household in the world whether that household is a human abode, nest, pit, hovel, cave, reef, riverbed, ocean floor, lair, or some other domicile. Work dealing with climate change always begins with the household. By studying up as Harding advocates Lahsen uncovers layers of scientific cultures that usually remain buried in the narrative of modern Western science defined by assumptions of exceptionalism, triumphalism and binary thinking. She ultimately seeks out a better form of science that is more responsive to the needs of those beings who dwell on the earth.

6.2 Paul Gilroy, Science, Race, and Nationalism

On the surface Paul Gilroy seems out of place when discussing matters of science and postcolonialism. Gilroy, a sociologist and leader of the now defunct Birmingham School of Cultural Studies writes about matters of race, nationalism, and postcolonial thought. Yet what I want to demonstrate is that sprinkled within his writings are distinct references to the natural sciences and how these disciplines connect to (post)colonialism. Gilroy is interested in science for two reasons. The first deals with colonialism and how science was and continues to be used to rationalize the existence of a racial hierarchy fueled by notions of national superiority and ethnic exceptionalism, and the second deals with the rise of new sciences and what he often refers to as the nano-technologies, other times as molecular biology, and still others as the DNA revolution. No matter which term he chooses, he points to the same concern: the remaking of the human body and the redefinition of what and, more importantly, who is referred to as human.

In his recent edited book, *Conflicting Humanities*, with Rosi Braidotti (2016, 98) in honor of Edward Said, Gilroy writes: "Belligerence and nationalism are vanquished through a variety of intellectual labour in which human curiosity and the attachment to culture surpass loyalty either to the national state or to the mechanisms of absolute ethnicity with which one finds oneself entangled." When Gilroy writes of belligerence he is referring to at least the last 200 years in the Western world in which a hierarchy of dominance was invented around concepts of race and ethnicity. This belligerence created a system rationalizing and justifying the enslavement of African, Asian, and American peoples. Part of this rationalization was the rise of scientific racism that became a hallmark of nineteenth and twentieth century thinking. Connected to notions of racial superiority and ethnic exceptionalism was the rise of nationalism in which Western European nations used science and technology to justify the dominance of other peoples throughout the world. Gilroy recognizes that politically, culturally, and economically, colonialism is in decline, but he also recognizes that it is not dead and can very easily reemerge as the dominate form of thinking in the world. His goal is to create ways to disconnect racism, ethnic absolutism, and nationalism from the destinies of individuals and allow all individuals to flourish in any creative way they feel necessary. Gilroy believes in order for this to happen science must be disconnected from race, ethnicity, and nationalism and the new sciences cannot be used to yet again limit the abilities and possibilities of peoples who have traditionally been oppressed and enslaved.

Gilroy views matters of race in a slightly unique manner than most scholars. In *After Empire: Melancholia or Convivial Culture?* Gilroy (2004, 42) writes "for me, 'race' refers primarily to an impersonal, discursive arrangement, the brutal result of the raciological ordering of the world, not its cause." In other words, race does not come first as scientific racists tried to demonstrate in the nineteenth and twentieth century in order to justify slavery and later segregation, immigration policies, and I.Q testing. Instead it is a raciology, or the notion that some groups of peoples defined later by racial and ethnic hierarchical classifications who seek to codify and camouflage their prejudices and stereotypes as reality and truth, that cause the creation of racial differences. It is important to note that Gilroy is not allying himself here with someone like Clarence Thomas who believes that race can magically be wished away or even with liberals who might believe that race can be overcome without major political, cultural, educational, and economic structural changes. For Gilroy a radical reordering of the world is needed in order to overcome centuries of inequality constructed around race. His point is that the racism, a notion of national and ethnic superiority of European peoples came first before the existence of notions of race. Race had to be invented in order to justify European dominance over the world, just as class had to be invented to rationalize the dominance of certain groups of people within Europe to justify their reign of power over the majority. In another book, *Against Race: Imagining Political Culture beyond the Color Line*, Gilroy (2000) warns that the "first task it to suggest that the demise of 'race' is not something to be feared. Even this may be a hard argument to win. On the other hand, the beneficiaries of racial hierarchy do not want to give up their privileges." For Gilroy a post racial society is a postcolonial world. The expectation should not be that those states and peoples who have constructed their authority and power in a racial, ethnic, and national world will concede their power easily. There is no reason that these powerful forces can regroup and reconstruct a rationalization justifying their dominance in a post-racial, post-ethnic, post-national world. In fact, it is not a hard argument to make and say this is exactly what has happened with the rise and dominance of post-national corporations who wield their power over peoples not within national borders but through international courts, international trade agreements, and international institutions such as the World Bank and the International Monetary Fund. In spite of these in-roads by multinational corporations, the near future of a post-racial, post-ethnic, post-national, and postcolonial world is still unwritten. The shackles of a racialized, ethnically absolutist, and nationalistic mentally have been barely removed, and in many places they still remain. It is the uncertainty of the moment that Gilroy believes makes it an important time to construct a new way of thinking about what it means to be a human being. This is where the importance of science plays an important role in his thinking.

In *Against Race*, Gilroy (2000, 15) warns: "Any inventory of the elements that constitute this crisis of raciology must make special mention of gene-oriented or genomic constructions of 'race.' Their distance from the older versions of race-thinking that were produced in the eighteenth and nineteenth centuries as the relationship between human beings and nature is constructed by the impact of the DNA revolution and of the technological developments that have energized it....we must try to take possession of that profound transformation and somehow set it to work against the tainted logic that produced it." For Gilroy the scientific racism that justified economic exploitation through colonization and slavery and rationalized the construction of a racial and ethnic hierarchical society based on false notions of biology and blood has been exposed as ideological and unscientific but the logic that undergirded notions of racial and ethnic superiority remain lethal ideologies. At any time these ideologies can potentially return in full force within and through the rise of new sciences thereby instituting a new era in which science is used to reconfirm or rationalize old racial and ethnic hierarchies. The problem with Gilroy is in his critique of the current state of postcolonial affairs is he does not give specifics when he is referring to matters of science. He does, however, offer clues.

One of the more important questions for Gilroy throughout his scholarly career is the question of who in humanity counts. Molecular biology, stem cell research, and genetics raise important and new concerns around the question of who counts. In the colonial world dominated by European nations and the United States, the colonialized were infrahuman or less than human. Whether they were "slaves" from West Africa, "primitive savages" from the Western hemisphere, or "orientals" from the East, they were constructed as infrahuman. With the emergence of the postcolonial after World War Two this question remains a contentious issue. Who counts for instance in India now that it is a sovereign nation-state? Do the so-called "untouchables," women, or the villagers mentioned by Abraham above count? What about the poor in Mexico City, Sal Paulo, Beijing? What about First Nations Peoples in Canada or the United States? Science matters in Paul Gilroy's thought because it offers a new twist in the question of who counts. Again in *Against Race*, Gilroy (2000, 20) suggests "the cultivation of cells outside of the body for commercial and other purposes is an epoch-making shift that requires a comprehensive rethinking of the ways we understand and analyze our vulnerable humanity." Just as fast as various peoples throughout the world earned at least a semblance of being recognized as human after World War Two, Gilroy warns that it can be removed even faster because of developments within sciences. Gilroy is not suggesting that science is inherently a tool of the dominant, but he is warning that science has been historically used in that way. Given the past of colonial dominance and the oppression of the majority of people in the world in the name of empire, racial purity, ethnic absolutism, and national superiority and the tendency of the natural sciences to be used to rationalize these discourses of dominance, it is important to grasp the impact the new sciences will have in defining who is human and who will be referred to as infrahuman or even worse as a commodity for human consumption. It is not a coincidence that Gilroy refers to the cultivation of cells outside of the body. Stem cell research is moving at a fast pace and issues so far regarding how they will be used and who shall gain access to anything productive from their development are barely debated. So far the so-called 'free market" has dictated how they are being developed and this path historically means only those with power and a disposable income will have access and those who do not will merely be erased from consciousness or rationalized as unfortunate souls who either do not have the need for access to stem cell research or lack the means to gain it. They will be labeled again, as in colonialism, infrahuman. This possible scenario dealing with stem cells and other commercialized forms of scientific development is already playing itself out in regards to organ transplants. Throughout the world the most vulnerable are preyed upon by underground kidney donor programs. The poor throughout the world are offered tens of thousands of dollars to donate one kidney and the affluent throughout the world hop on a plane for a transplant vacation. The recipient is defined as worthy of life and the "donor" is a commodity on the world organ market. If these matters of science are not addressed and arguments for the sanctity of all human and non-human animal life are not created, a return to colonial mentalities, policies, and politics will emerge within the realm of possible realities, and science will be a major engine in rationalizing its return. Within the context of this tragic possible future, Gilroy (2004, 41) asks this important question: "What are the raciological differences becoming next, in a world where our understanding of humanity has been irrevocably reshaped by genomics, biotechnologies and self-conscious biocolonialism?" Will the future bring a world of scientific developments debunking the myth of race and institutionalizing all humans as equal or will a new form of the Ancien Regime emerge using science to justify the sanctity of life based on race, ethnicity, and nationality while turning invisible the infrahumans who fall outside the borders of these human constructs? This is Gilroy's challenge to us.

6.3   Sylvia Wynter, the Poetics of Science, and the Rise of Man3

While Harding presents a comprehensive approach to postcolonialism and science and Paul Gilroy warns of possible consequences for ignoring the impact of science on postcolonial thought, it is Sylvia Wynter's thought that is most intriguing. Yet, when Wynter's project and thought are commented on, science is rarely mentioned. What she proposes is a rethinking of science. Influenced by Aimè Cèsaire and Franz Fanon, Wynter simultaneously acknowledges the importance of science in postcolonial thought and its gaping silences. Wynter proposes what Walter Mignolo (2015, 116) has referred to as a decolonial scientia. Scientia is not a rejection of an empirical approach to science developed by Bacon, Boyle, Copernicus, Galileo, Newton, and others but it recognizes that science is very limited in what it can claim as truth and what it can reveal about a reality. Decolonial scientia is the science "needed not simply for progress or development but for liberating the actual and future victims of knowledge for progress or development" (Mignolo, 2015, 116). Following Cèsaire, Wynter wishes to reunite science with poetry. It is poetry that will bring life back to the sciences, it so desperately needs. For Cèsaire (1996, 134), "Poetic Knowledge is born in the great silence of scientific knowledge….science offers…a perspective on the world. But a summary one. One that is superficial…. Scientific knowledge enumerates, measures, classifies and kills." Wynter not only draws from postcolonial scholars such as Cèsaire and Fanon but also from scientists such as Heinz Pagels who "suggests that…a reinvented study of letters would have to be based on the erasing of the barrier between the natural sciences and the humanities, as the condition of making our 'narratively constructed worlds and their orders of feeling and belief' subject to 'scientific description in a new way'" (Wynter 1997, 162). This section looks at how Wynter brings poetry and science together in order to create an alternative vision, a new way that does not dehumanize large swaths of people in the name of progress and development.

Similar to Gilroy, Wynter is interested in uncovering how non-European peoples were classified and constructed as infrahuman. In her work Wynter constructs a genealogy that covers the history of humanism, the power of religion, secular European politics, Darwinian science, and the rise of postcolonial thought. Her genealogical studies take her back to Columbus where we see the emergence of what she refers to as Man1. As Europeans made their way throughout the coasts of Africa and the Western hemisphere they not only sailed with economic and scientific agendas in mind but also beliefs of who they were and what they embodied. As soon as they encountered different peoples from Africa and the "new" world these beliefs began to shape how these peoples were to fit into the worldview of Europeans. Katherine McKittrock describes Man1 as an invention of the Renaissance and a morphed version of Medieval Christian man, homo religiousus. Man1 is "studia humanitatis as homo politicus and therefore differentiated but not wholly separate from the homo religiousus conception of human" (Wynter and McKittrock 2015, 10). When Man1 encountered other forms of humanity, he began to construct a hierarchical system to explain these new encounters. If Man1 were sanctioned by God as the image of God but Man1 was free from the sanctions of God's church as an autonomous individual free to live out his wishes over God's dominion how is Man1 to interpret his new encounters? Are these "newly discovered" peoples humans, can they be grouped with the men of Man1? What emerged from these encounters were religious doctrine, imperial policies, and cultural interpretations of Man1's interpretation of these new encounters. The path selected was to rationalize the treatment of these new peoples as less than human, as not Man, not European enough. But were they at least human? Wynter demonstrates through the work of Shakespeare that this path can be seen emerging. In an interview with David Scott (2000, 179), Wynter notes "when you read Shakespeare, it is not...the invention of the human that is at issue. It is the invention of the first de-godded Man that we are seeing in Prospero and in Caliban who is his other." This de-godded man is simultaneously the Renaissance Man freed from the shackles of a dogmatic church but still sanctioned by "God" to hold dominion over the world and the de-godded other who is not sanctioned by God and possibly completely disconnected from God. This other de-godded Caliban figure soon Wynter notes will pick up the labels of Indians and Negro/a. Those who receive the label of "Indian" will be "portrayed as the very acme of the savage, irrational other, the 'negroes'...represented as its most extreme form and as the ostensible missing link between rational humans and irrational animals" (Wynter 2003, 266). This debate is not just found in Shakespeare's *The Tempest*, it is also found in Roman Catholic Church doctrine. If these "Indians" and "Negroes" are found to be human then they cannot be enslaved and they must be given the chance to convert to Christianity since they never received a chance to hear the word of Christ. However, if they were to hear the word of Christ and reject it then they could be enslaved as heathens ("Enemies of Christ/Refusers of Christ) or if they were deemed to be not human then they also could be enslaved. What emerges from these debates Wynter (2003, 296) suggests is the first modern concept of race,

The councils of jurists/theologians that King Ferdinand set up for this purpose [to define who was human] had come up with a formula that, adopted from *The Politics of Aristotle*, would not only enable the master trope of Nature (seen as God's agent on Earth) to take the latter's [God] authoritative place, but would also effect a shift from the Enemies-of Christ/ Christ Refusers system of classification to a new and even more powerfully legitimating one. It was here that the modern phenomenon of race, as a new, extrahumanly determined classificatory principle and mechanism of domination...was first invented...For the indigenous peoples of the New World, together with the mass-enslaved peoples of Africa, were now to be reclassified as 'irrational'... 'savage' Indians and as 'subrational Negroes....

A new notion of the world and 'idea of order' was being mapped now, no longer upon the physical cosmos [God]...Instead, the projected 'space of otherness' was now to be mapped on phenotypical and religio-cultural differences between human variations and/or population groups while the new idea of order was not to be defined in terms of degrees of rational perfection/imperfection...as that of the 'law of nature'

This reconfiguring of defining who is human wrests the authority to do so from the Church and places it into the realm of Man1, homo politicus, the secular monarchical West and sets the stage for the emergence of Man2. There is no longer a need to use a god to rationalize a classificatory systems as to who is human and who is not. Nature now can play that role. There is no need for a cosmic being to make decisions for humans when the new anthropomorphized being, Nature, can now do it and all the laws of humans can be ordered around it.

Recall, Gilroy's idea that it is a raciology that is needed before the concept of race can be invented. This is exactly what Wynter is suggesting here, but she takes Gilroy's notion further. For Wynter the notion of who is human and who is not becomes a phenotypical decision leading to the emergence of scientific racism and the invention of race. For Wynter (Wynter and McKittrock 2015, 10) Man2 is "a figure based on the Western bourgeoisie's model of being human that has been articulated as, ...liberal monohumanism's *homo oeconimicus*...who practices, indeed, normalizes, accumulation in the name of (economic) freedom. Capital is thus projected as the indispensable, empirical, and metaphysical source of all human life, thus semantically activating the neurochemistry of our brain's opiate reward/ punishment system [Mill's Utilitarianism and neoclassical economists' rational man model or choice theory] to act accordingly!" The foil in the Man2 model is no longer "Indians" or "Negroes" but the disenfranchised, homeless, and poor. For Wynter (2003, 314) what emerges by the end of the nineteenth century is a redefinition of humanity, "Man (as embodied in Prospero) was redefined as optimally economic Man, at the same time as this Man was redefined by Darwin as a purely biological being whose origin,...was sited in evolution, with the human therefore existing in a line of pure continuity with all other organic forms of life." This redefinition of Man as purely biological, as well as economic, but not theological, leads us to our current scientific circumstances in which humanity is being defined by "the biological sciences and its...contemporary, dazzling triumphs...the cracking of the DNA code, the Human Genome Project, together with the utopian cum dystopian promises and possibilities of biotechnology" (Wynter 2003, 314).

From preMan1 to Man1 and through to Man2, Wynter argues that these inventions all represent a system of projection and human mythmaking or what she refers to as Mythoi. These myths are embodiments of the human ability both in science and the arts to invent stories of who Man is. This includes the myth that a cosmic being gave dominion of man over the earth or that a cosmic being gave European Man the authority to define who was human and who was not, it also includes the myths that Man is purely biological thereby setting the stage to construct the idea of race and encourage the invention of scientific racism. These myths all represent the human tendency to displace and project their invention onto other entities such as a god, nature, or science. It is this projection onto other entities that allows Man1 and 2 to erase, ignore, and forget that he is creating these myths and projecting them onto others in the first place. This erasing, ignoring, and forgetting of humanity's ability to invent embodies its greatest blind spot. It is this blindness that allows humans to justify the genocidal destruction of peoples and cultures, enslavement of others, and today to deny the poor any role in the defining of what humanity and life means. It is this blindness that prevents humanity to see it is autopoetic. It is at this point Wynter's constructs an alternative vision to the human inventions Man1 and Man2 as she brings science and poetry together. For Wynter "what 'we really have is a poverty-hunger-habitat-energy-trade-population-atmosphere-waste-resource problem' none of whose separate parts can be solved on their own. They all interact and are interconnected and thus, together, are constitutive of our species' now seemingly inescapable, hitherto unresolvable 'global problematique'" (Wynter and McKittrock 2015, 44). We are living a global, human created, economic, scientific, and cultural problem that she will later name as climate change caused by an inability to see that humans have created this problem not a god, nature, or pure science. This earthly and worldly threat requires a rethinking of Western traditions that invented the mythoi of Man1 and Man2 and new way of viewing humanity and the world with all its other inhabitants. Wynter's vision requires the abandonment of Man1 and Man2 and recognize that humans are not purely biological, therefore classifiable by race or phenotype but instead humans are "a hydid being, both bios and logos (or as I recently come to define it, bios and mythoi) Or, as Fanon says, Phylogeny, ontogeny and sociogeny, together, define what it is to be human" (Wynter and McKittrock 2015, 16). To be human is to be both scientific and poetic. But to be both, all humans have to be recognized as human and not rationalized away as "savage," "infrahuman," or "poor." Wynter's system of thought demands the recognition of all humanity not just "God's" chosen few or "the economy's" selected leaders, or "phenotypes'" superior races. If it is science that constructed Man2 and scientific racism, it is also science that plays a part in escaping the inhumanity of these human inventions. Wynter's alternative calls for *"the autopoiesis of being hybridly human"* which she coopts from "Maturana and Varela, who wrote the book *Autopoiesis and Cognition*" (Wynter and McKittrock 2015, 27). Maturana and Varela play an important role in Wynter's thinking because their concept of autopoiesis starts a revolution in cybernetics and numerous other fields within the sciences when they posit that it is not reality that imposes images upon the individual but it is the individual who constructs a reality within their minds and imposes it on an outer reality. Such a perspective provides Wynter the metaphor to describe how humans are hybrids and how they project their own mythoi upon the world and then call it "truth," "nature," "god," or "science." Naming a system of thought and institutions of religion, science, economics, and politics mythoi is only the beginning. Just because human systems are revealed to be autopoietic does not mean maladies will be cured and human inequalities disappear. Human inequalities are never guaranteed to disappear, there is an abundance of evidence in existence as I write to tell us otherwise. Wynter's hybrid human is a call for those who embrace postcolonial thought to enact this thought into an action that dismantles systems that enslave peoples and other non-human animals. This means a dismantling of neoliberal politics and neoclassical economics as well as the new form of preMan1 thinking, Protestant and Islamic fundamentalism, both forms of world-wide terrorism. It also requires an engagement with science and what I refer below as Man3 or a potentially new way emerging to redefine humans into mythical classifications based on notions of superiority and inferiority.

In his book *Life out of Sequence: A Data-Driven History of Bioinformatics*, Hallam Stevens (2013) conducted an ethnographical study of the academic field and business of bioinformatics. Literally studying up from human and non-human D.N.A. what he charts I believe is the potential for the formation of a Man3. Stevens (2013, 6) notes that "'bioinformatics' is used here as a label to describe [an] increasing entanglement of biology with computers…Bioinformatics does not have a straightforward definition;…it is a site of contestation about what biology will look like in the future: What forms of practice will it involve? How will knowledge be certified and authorized?…what roles will computers and the digital play in biological knowledge?" Once these basic definitions and questions are raised he takes the reader on an ethnographic trip on the rise of biogenetics as an industry and the role that data plays in defining what it means to establish a biological fact and ultimately what it means to be a human as well as what it means to be a non-human animal, and an organism. When we finish our tour Stevens raises some very important issues that relate to the issues raised by Wynter and the others covered in this chapter. He concludes that bioinformatics as a field is disappearing because the biological is disappearing into data whereby biology is becoming data based materiality. "Human genomes," Stevens (2013, 205) suggests, "are increasingly understood by comparing them with the genomes of the hundreds of other fully sequenced organisms—and not just organisms that are perceived to be similar to humans." Bioinformatics and the data it generates is making humans a new hybrid form of bios but not necessarily a logos or poetics Wynter referred to earlier. Part of this hybridity is not just the merging of human genome with the genomes of other organisms but also the merging of human materiality with computer data generation. As Stevens (2013, 206) notes "next-generation sequencing machines are making more and more data an increasingly taken-for-granted part of biology." We are becoming data, what Stevens (2013, 219) calls *Homo Statisticus* or "new 'individualities'…a human constructed from the statistical residue of his or her genome." I will add these new potential individualities are not just residue from our own genome but we are becoming the residue from other non-human genome and computer generated statistical data. Will this residue of a person include the poor, the jobless, the homeless Wynter writes about? Or will it be the rise of a new Man3 in which certain, statistically generated and constructed individuals, "races," "ethnicities," or "nationalities" will be mythologized as superior and rightfully, truthfully, scientifically, and religiously anointed the chosen, privileged few by "god," "nature," or "science"? How this new configuration of humans, and I will add non-human animals as well, might be and should be configured is a question curriculum scholars should take up, but this is hard to do when science matters are barely on the agenda.

References

Abraham, I. (2000). Postcolonial science, big science, and landscape. In R. Reid & S. Traweek (Eds.), *Doing science + culture* (pp. 49–70). New York: Routledge.

Anderson, W. (2002). Introduction: Postcolonial technoscience. *Social Studies of Science, 32*(5/6), 643–658.

Braidott, R., & Gilroy, P. (Eds.). (2016). *Conflicting humanities*. London: Bloomsbury Press.

Cèsaire, A. (1996). Poetry and knowledge. In M. Richardson & K. Fijalkowski (Eds.), *Refusal of the shadow: Surrealism and the Caribbean* (pp. 134–146). London: Verso Press.

Chambers, D., & Gillespie, R. (2000). Locality in the history of science: Colonial science, technoscience, and indigenous knowledge. *Osiris, 15*, 221–240.

Cook, H. (2007). *Matters of exchange: Commerce, medicine, and science in the Dutch Golden Age*. New Haven: Yale University Press.

Gilroy, P. (2000). *Against race: Imagining political culture beyond the color line*. Cambridge, MA: The Belknap Press of Harvard University Press.

Gilroy, P. (2004). *After empire: Melancholia or convivial culture?* London: Routledge Press.

Harding, S. (Ed.). (1993). *The 'racial' economy of science: Towards a democratic future*. Bloomington: Indiana University Press.

Harding, S. (1998). *Is science multicultural?: Postcolonialism, feminisms, and epistemologies*. Bloomington: Indiana University Press.

Harding, S. (2008). *Sciences from below: Feminisms, postcolonialities, and modernities*. Durham: Duke University Press.

Lahsen, M. (2005). Seductive Simulations? Uncertainty Distribution Around Climate Models. *Social Studies of Science, 35*(6), 895–922.

Lahsen, M. (2008). Experiences of Modernity in the Greenhouse: A Cultural Analysis of a Physicist 'trio' supporting the Backlash against Global Warming. *Global Environmental Change, 18*, 204–219.

MacLeod, R. (2000). Introduction. *Osiris, 15*, 1–13.

McKittrick, K. (2015). *Sylvia Wynter: On being human as praxis*. Durham: Duke University Press.

Mignolo, W. (2015). Sylvia Wynter: What does it mean to be human? In K. McKittrick (Ed.), *Sylvia Wynter: On being human as praxis* (pp. 106–123). Durham: Duke University Press.

Schiebinger, L. (2004). *Plants and empire: Colonial bioprospecting in the Atlantic world*. Cambridge, MA: Harvard University Press.

Scott, D. (2000). The re-enchantment of humanism: An interview with Sylvia Wynter. *Small Axe, 8*(September), 119–207.

Seth, S. (2009). Putting knowledge in its place: Science, colonialism, and the postcolonial. *Postcolonial Studies, 12*(4), 373–388.

Stevens, H. (2013). *Life out of sequence: A data-driven history of bioinformatics*. Chicago: University of Chicago Press.

Wynter, S. (1997). Columbus, the ocean blue, and fables that stir the mind: To reinvent the study of letters. In B. Cowan & J. Humphries (Eds.), *Poetics of the Americas: Race, founding, and textuality* (pp. 141–163). Baton Rouge: Louisiana State University Press.

Wynter, S. (2003). Unsettling the coloniality of being/power/truth/freedom: Towards the human, after man, its overrepresentation – an argument. *CR: The New Centennial Review, 3*(3), 257–337.

Wynter, S., & McKitttrick, K. (2015). Unparelled catastrophe for our species? Or, to give human-ness a different future: Conversations. In K. McKittrick (Ed.), *Sylvia Wynter: On being human as praxis* (pp. 9–89). Durham: Duke University Press.

Chapter 7
Working Our Way Back: Colonial Science in Light of Postcolonial Thought

In the Previous chapter I dealt with the theories crafted by Harding, Gilroy, and Wynter as they are connected to science matters. In this chapter I want to extend their thoughts to the history of colonial science to look at the rise of colonialism and colonial science to its present day and how science can be reinterpreted and understood differently. In particular I want to look at the rise of botany as a natural intellectual emergence from the rise of colonial empires, the centuries old topic of bioprospecting and its companion biopiracy that continues to this day, and the history of the rice of Africa and the American colonies. The rise of botany represents an important example of locality mentioned in the previous chapter, bioprospecting embodies Gilroy's notion of raciology and the challenges of a post-racial, post-ethnic, and post-nation- state world can be seen in bioprospecting, and rice serves as an example of Harding's studying up and Wynter's Mythoi in colonial thinking. In the case of botany I will rely on Daniela Bleichmar's (2012) work *Visible Empire: Botanical Expeditions & Visual Culture in the Hispanic Enlightenment*, for bioprospecting Londa Schiebinger's (2004) work *Plants and Empire: Colonial Bioprospecting in the Atlantic World*, and for rice Judith Carney's (2001) *Black Rice: The African Origins of Rice Cultivation in the Americas*. While the work of these scholars will constitute the center and most detailed part of this chapter, the work of other scholars will also be used to highlight the changes, as a result of postcolonial thought, in understanding the links between empires, economics, and the sciences as well as between cultures, societies, and the sciences. Because of postcolonial thought, the very idea of who can do science is being challenged and rethought.

The issue of who can do legitimate science is part of the legacy of colonial science and the mentalities it tried to cement in the minds of everyone as universal and natural. The assumption that Euro-centric science is superior, although challenged, still lingers as we can witness in how non-Western science is treated within international intellectual communities. However there are signs of change emerging and it is important to take note of how mentalities concerning who can do science are being re-thought.

© Springer International Publishing AG, part of Springer Nature 2018
J. A. Weaver, *Science, Democracy, and Curriculum Studies*, Critical Studies of Education 8, https://doi.org/10.1007/978-3-319-93840-0_7

With the rise of colonialism, Eurocentrism developed. As Dipesh Chakrabarty has established from the rise of colonialism in the fifteenth century and through the rise of postcolonial challenges in the 1950s and 1960s, Eurocentrism has persisted as the role and abilities of non-Western nations in the postcolonial world were debated. European intellectuals noted that for non-Western nations to emerge as successful as Western nations they will have to "catch up". The non-Western nations will have to modernize, adopt Western notions of democracy, public and private space, and other entrenched ideals that embody Western concepts of development and civilization. As a result of these constructs and lists of requirements as to what it meant to be developed or civilized, non-Western nations always found themselves, no matter what the topic, ideology, or time period, behind. Chakrabarty (2002) for instance uses the example of the work of noted Marxist historian Eric Hobsbawm to stress this point. According to Hobsbawm, the problem with a nation such as India and its struggle for development and modernization was the traditions of the Indian masses were too premodern, therefore the nation of India could not emerge as part of the developed world. What needed to happen was the masses of India would have to transform themselves from peasants to industrialized workers. Once the masses of India develop a class consciousness then they can become political and demand their rights. Embedded in Hobsbawm's notion of modernism and development is the notion that Europe was the universal standard to judge who was a developed nation and who was still too primitive and traditional to be part of this select group. No other alternative paths to become independent peoples were imagined or seen as possible from the perspective of Eurocentrism because non-Western peoples had no theory, history, or political means to create alternative paths. Yet, this is exactly what happened with the case of India and other nations. As Chakrabarty (2002, 19) notes the "peasant did not have to undergo a historical mutation into the industrial worker in order to become the citizen-subject of the nation." What emerged was a hybrid in India: A modern nation-state in which Indian traditions remained, with all their problems and concerns while a viable and stable democracy developed.

A similar Eurocentrism dominated the thinking of science. Eurocentric science insists that the way Western science is done is universal and absolute. The use of set protocols, abstract thinking, and universal truth claims are needed for any work to be labeled as legitimate science. Anything else was illegitimate and suspect. The historians of science mentioned in this chapter challenge Eurocentrism and offer alternative ways to see how science is done. From their work what emerges are different perspectives of how science was, is, and can be done that benefits not only multinational corporations and Western nations that often do the work but all parties associated with and effected by the research. One of the aspects of the work of Bleichmar, Schiebinger, and Carney is that the research developed by so-called Western scientific ways were never just influenced by Eurocentric traditions and ideas. The erasure of non-Western traditions from post-Columbian era science was one of the major myths defining colonial science.

7.1 Locality and Colonial Botany

If there is ever a clear case of the connection between science and colonial empires it is botany. Botany may have emerged without the assistance of empires as a scientific field but it would never have developed as a major field of study. As the Spanish, Portuguese, British, Dutch, French, and even the minor Swedish empires developed, the colonial lands were immediately envisioned as economic treasure troves of exotic spices and plants exploitable for enriching monarchical coffers. While the known spices of pre-Columbian era were still sought, empires did not know what new spices might exist and what new species might become new profit making commodities. For instance, when the Spanish were introduced to tobacco and chocolate by the Western Hemispheric peoples, there was no clear vision of what these new experiences in taste meant. Tobacco was used in sacred ceremonies and often used for pain relief. Yet as Marcy Norton (2008, 56) notes tobacco appears in the influential work of Gonzalo Fernandez de Oviedo under a chapter "devoted to the 'crimes and abominable customs and rites' of the indigenous people of Hispaniola." Such testimony helped establish tobacco as a plant of the devil. Eventually, these views were challenged. For instance in 1571 Nicholas Monardes (Norton 2008, 116) reported good medical news "For asthmatics, 'taken in smoke by the mouth [tobacco] expels chest material.'" Just as Wynter reports in her work that theologians debated whether peoples from Africa and the "new" world were human or not, botanists, medical doctors, literary authorities, and philosophers debated the value of tobacco. The same scenario played out with chocolate. In order to know which plants and their byproducts would be profitable for empires to exploit and claim as their own a whole structure had to be created. By the end of the eighteenth century every empire had a centralized system of botanical gardens headed by cabinet botanists who managed the collection, organized experiments with plants and trees in the gardens, and constructed a whole network of travelling botanists who collected the plants and trees and reported back to the central intellectual leaders whose gardens were located in Madrid, London, Paris, Uppsala, and Berlin. The cabinet botanists were the eyes of authority. They established the importance of a plant, tree, or animal for science and empire. The travelling botanists, however, were important for these figures of intellectual authority. As Bleichmar (2012, 32) notes the Swedish botanist Karl Linneaus referred to his vast array of travelling botanists as his "'apostles.'" They were the eyes that allowed the cabinet botanists to establish scientific and economic "fact" without leaving the hubs of power and scientific work. The travelling botanists spread the word as to what was accepted as important epistemologically and economically. From a Eurocentric perspective, this is all that mattered in order to create botany as a science. All other dimensions of the making of botany as a science were unimportant and erased. This is where Bleichmar's work enters. In *Visible Empire* she recognizes the role that cabinet and travelling botanists play in creating knowledge, but there were other players including artists in the making of this knowledge and locality was important in making decisions as to which plants, trees, or animals would be uprooted or captured and shipped back to Europe so that scientific and economic decisions could be made as to the importance of a plant or tree. "My analysis…refutes center-periphery models of knowledge production" Bleichmar (2012, 14) asserts, "demonstrating that the Spanish empire functioned instead as a network with multiple nodes and competing interests." What these local sites of scientific production decided were important examples of plants, trees, and animals, and how effective the local artists, both indigenous and European, were in capturing the essential aspects of a species, dictated how scientific knowledge and economic policies were established in European centers. What emerges from Bleichmar's work is a network in which lines of communication and the flow of knowledge do not move only from center to periphery but they flow from local areas of knowledge generation such as Mexico City, Lima, and Bogotá where travelling botanists set up their headquarters to begin collecting species and learn the local terrain. Bleichmar's challenge to the idea of scientific networks working from a center-periphery relationship includes the work of Latour. "According to Bruno Latour, the production and circulation of knowledge in European expeditions operated cyclically and iteratively. Data were useless 'out there'; a voyage's success depended on mobilizing information to the metropole, the center of the network" (Bleichmar 2012, 141). This approach places all the importance of defining knowledge at the center while the whole process of deciding, finding, creating, collecting, and shipping, remain on the margins as pre-knowledge and eventually erased from the narrative. "Although Europe always remained the ultimate frame of reference as the source of funding, prestige and significance," Bleichmar (2012, 141–142) concedes, "Spanish and Creole naturalists were involved in both metropolitan and local agendas [of collection as were local] institutions of higher learning, printing presses, important private libraries [often of local experts of indigenous plants, trees, and animals] and active intellectual and artistic communities, these men found themselves working very much in the thick of things, not on the edge."

It was this local collection of government leaders, printers, artists, intellectuals, and private citizens who enabled the construction of a network to create what Bleichmar refers to as, the "visual epistemology." It is this visible epistemology that allows the cabinet botanists to make any decisions about the scientific and economic viability of a species captured in paint. The creation of a visible epistemology is a matter of local knowledge, understanding of practical usage already in place in the indigenous communities, and important expertise a center-periphery model fails to acknowledge. "Making visible or visualizing are not the same as seeing" Bleichmar (2012, 38) insists. "My point is that making visible—a process that involved not only the final viewing of an image but also the acts of observation and representation that yielded it—had both pragmatic and symbolic dimensions and was widely understood as an integral part of the process of producing knowledge and enacting governance." What emerges from this idea of a visible epistemology is that before a cabinet botanists could make a determination if a plant or tree is unique and should be shipped to a European garden and before a court economist or policy maker could make a decision about the profitability of a new commodity, it was the local scientific actors who were making all of this possible. The central makers of scientific fact and economic profit are dependent on the local knowledge, artistic talent, and imaginative abilities of the various actors at the local level that moved well beyond the travelling botanist. If the artist does not visualize the important parts clearly enough and know what makes a certain species of plant, tree, or animals unique, the power brokers back in Europe could not even think about making a determination of its importance scientifically or economically. "Visual epistemology," Bleichmar (2012, 46) notes, "bridged distances, transporting nature in multiple incarnations—as picture, specimen, or text—so that it became available to naturalists in different locations, allowing them to see at a distance and to collaborate with one another." Rather than being on the periphery as a bit player in a scientific process the local artists, intellectuals, printers, and community leaders were the eyes that made seeing possible, they were the information that made communication necessary and possible, and they were the ones who made botany possible. They made the establishment of scientific fact possible.

7.2 Bioprospecting, Agnotology, and Raciology

While Bleichmar focuses on the construction of knowledge and role of creating Botanical knowledge, Londa Schiebinger is interested in Botany as bioprospecting (some say biopiracy and I will use that word as well to mean the same thing. Bioprospecting can be seen as another example of the role locality plays in the making of science but here I want to focus on bioprospecting as a process of rationalizing the superiority of Europeans and inferiority of all others including European women and the function bioprospecting plays in privileging certain forms of knowledge while erasing and ignoring other forms of scientific knowledge. The word agnotology is a word Robert Proctor asked a colleague, Iain Boal, to create to describe the kind of historical work Proctor was doing in regards to the cultivation of ignorance. As Proctor (2008, 27) notes the word agnotology draws from two Greek words: "agnoia, meaning 'want of perception or knowledge,' and agnosia, meaning a state of ignorance or not knowing, both from gnosis…meaning 'knowledge.'" Schiebinger, like Proctor, wishes to use this term not to mean that people just do not know something or that they willfully do not wish to know something. These forms of ignorance remain but that is not what interests Schiebinger. Agnotology is ignorance "as something that is made, maintained, and manipulated by means of certain arts and sciences…that certain people don't want you to know certain things, or will actively work to organize doubt or uncertainty or misinformation to help maintain (your) ignorance" (Proctor 2008, 8). If we look at the history of colonial science as a form agnotology, then bioprospecting is not only just the search for biological or botanical knowledge but it is withholding of certain knowledge deemed as unimportant or dangerous to certain populations. It can also mean the willful erasure of certain knowledge in order to keep from other people or, in the case of postcolonial thought, to prevent the acknowledgment of certain peoples as humans. It is these parts of bioprospecting I wish to focus on in this section in order to eventually demonstrate how we can connect the history of science to Paul Gilroy's postcolonial thought, his notion of raciology and how it works through science matters.

Often when bioprospecting is a topic of discussion, the main issue is the finding, naming, classifying, and exploiting of plants and animals to develop new agricultural or pharmaceutical commodities to sell either by stealing the plant or tree from another territory, people, or government thereby erasing the historical connections and legal rights others may have on claiming the commodity as theirs or through protections such as patents, international laws, and trade agreements in order to claim originality and ownership. This is certainly still part of the story of postcolonialism and current multinational, global capitalism. However, Londa Schiebinger offers another way of looking at bioprospecting that includes the role gender and race play in establishing some plant or tree as a valuable specie, and manipulated ignorance and secrecy plays in defining what species are important to cultivate and understand. Schiebinger in her work looks at the peacock flower and its ability to act as an abortifacient. The peacock flower as an abortifacient can be seen from at least two different angles. It can be seen as something that African female slaves wish to keep from Western medical doctors and plantation owners, and it can be seen as a home remedy Western medical doctors wish to keep from European midwives and women.

Botany like all sciences is a gendered space. When bioprospecting and the development of botany emerged as a part of science, males dominated the field both as cabinet botanists and travelling botanists. Gender shaped what botanists saw when they were travelling just as much as economics shaped the rationale for going on excursions in the first place. Male botanists were interested in plants that could increase imperial treasuries and expand scientific knowledge as it was defined by other males. Botanists sought out organic remedies for the many ailments Europeans suffered from the fifteenth century to the present; ailments that afflicted both men and women. But when it came to female specific needs botanists had trouble seeing the utility of a plant or tree. It should be no surprise then that it was one of the few female travelling botanists, Maria Sibylla Merian, who was the first European to recognize the peacock flower for its medicinal value? When Merian returned to Europe and published her reports she included a discussion of the peacock flower, but the medicinal value of the plant never made it into the medical commentary of the times. Schiebinger (2004, 153) notes "although the peacock flower was taken many times into Europe, the knowledge of its use as an abortifacient did not transfer to Europe." There were numerous reasons why the peacock flower never emerged as a part of official medical knowledge in Europe. First, Merian as a woman was not a part of the official botanist field. Most botanists at the time were medical doctors therefore trained at the university; something denied Merian. Second, most women's health concerns were taken care of by midwives and midwives were not seen either as part of the medical community nor part of the official knowledge networks. Instead they were viewed as healers, and their approach to medicine was not viewed as legitimate knowledge (Schiebinger 2004, 96). Third, by the time Merian arrived on the scene of botany doctors were beginning to usurp the power of midwives and control their activity more intensely. All of this does not mean that medical doctors and government officials were not aware of the medical potential of the peacock flower. They were, but, in the spirit of agnotology, they wished to prevent midwives and women from gaining any knowledge of this flower. Like any topic concerning the flow of information, preventing midwives from spreading the news about the peacock flower or more traditional European abortifacients such as Savin became an impossible task which did not result in the relaxing of laws against the use of abortifacients but merely began a phase that continues today; the control of women's health by men. Schiebinger (2004, 122) demonstrates that control of midwives began as early as the 1500s and "prohibited them from deliberately causing abortions." Whenever a midwife knew a woman was pregnant she had to report the pregnancy to a medical doctor who would rarely play any role in the care of the mother during pregnancy nor partake in the birthing process, but if a mother did not come to term with the fetus the medical doctor would know an abortion might have occurred which would then begin an official inquiry. By banning women from the university, denying the validity of their medical knowledge, controlling midwives use of abortifacients and the dissemination of their knowledge, medical doctors and government officials deliberately used a strategy of agnotology to deny women the best possible health care available at the time.

While knowledge of the peacock flower and other abortifacients were denied official status and usage in Europe, African women knew of the peacock flower's medicinal power and used it to their advantage. As Schiebinger (2004, 128) reveals "Abortion, especially among slave populations, was not a matter of private conscience or 'family planning' as we might think of it today; it was rather a part of the colonial struggle of victors against vanquished and a matter of economy and state." When slavery became common practice to provide workers to develop plantations in South America, the Caribbean, and North America, Africans were forced to make the journey across the middle passage. They were able to bring a few things along with them, among those items, as I will show later in this chapter, were rice and abortifacients such as Okra and the peacock flower. The abortifacients became part of the colonial struggle. "Abortion," Schiebinger (2004, 131) notes, "then, was only one type of resistance among many, and a number of contemporary observers saw it in such terms." While Europeans debated whether Africans were humans who needed to be converted or not or subhumans who were natural enemies of god and therefore able to be enslaved, those Africans who were enslaved knew the horrors of the practice and were prepared to resist anyway they could. Abortifacients gave women assurances that their children, as a result of rape or consent, from the seed of African or European men, would not suffer "the horrors of bondage" (Schiebinger 2004, 145). What is interesting about this battle over life in African women is the role of knowledge was often reversed. Colonial plantation owners and officials knew abortions were happening at a higher rate than normal but they often did not know how. If they did know by which means abortions were being induced, they then did not know how the concoction was being used. Abortifacients then became a reversed case of agnotology. The African women, as is also the case with indigenous peoples of the Western hemisphere who used local flora as abortifacients, withheld medical knowledge from the colonialists in order to control the fate of their own offspring. Abortion and the knowledge of when to induce one remained a constant struggle throughout the existence of slavery in the Western hemisphere.

What these cases of abortifacients and the use or non-use of medical knowledge to control women's bodies demonstrate are other ways of thinking about Gilroy's notion of raciology and the construction of race to justify a system of inequality. Both European and African women were erased as human beings who had the capacity to construct and utilize human knowledge. From the perspective of Western men they were both subhumans and as a result a system had to be created to rationalize why women in Europe were not capable of attending universities and becoming medical doctors, and at the same time a system had to be invented to construct African women as subhumans only worthy of slavery. Bioprospecting served as a means to justifying this system of erasure but with very different results. When it came to European women the limits of bioprospecting to search for plants and trees that served the purposes of what men defined as knowledge and therefore important resulted in the further control and subservience of midwives under the control of male medical doctors. In the case of African slaves, bioprospecting limited the knowledge of Western colonialists regarding the local ways of birth control thereby limiting their abilities to control the forms of resistance women used against the colonial powers. In the first case bioprospecting promoted a form of agnotology that helped control women and in the other case bioprospecting limited the knowledge of Europeans.

Concerns with bioprospecting remain, as does raciology, but the circumstances are very different. A system rationalizing the erasure of Africans or indigenous peoples as humans is no longer accepted as a means to exploit the flora and fauna or the people involved. International trade agreements, multinational corporate drug protocol trials, and scrubbing data are the preferred means of rationalization. In many cases international trade agreements allow corporations to enter into any nation to develop a potentially profitable product. John Merson (2000) in his work uses the Neem berry as an example. In an effort to create less harmful insecticides than those created by chemical companies, numerous bioprospecting multinational corporations have sought alternative forms of repellants. The Neem berry acted as an insect repellant but did not have any harmful side effects to crops, the environment, or people. The USA based company W.R. Grace began production of the new product that used the Neem berry and a one billion dollar industry emerged. Merson (2000, p. 289) noted that while trade agreements in the 1990s offered protections to the nation of India against corporate exploitation "no legal mechanism recognized the collective intellectual property interests of traditional users." The rights of the indigenous people to the use of the plant were denied and, according to the trade agreement, given to the nation-state. Trade agreements and patent rights, never a topic of discussion when colonial powers rationalized the exploitation of plants, trees, and animals as well as indigenous peoples now utilize these legal means to enter into agreements with nation-states to ensure their legal rights to profit making while the needs and cultures of indigenous peoples continue to be usurped and overlooked.

Bioprospecting now has grown to not only include plants, trees, and animals but humans as well. Anthropologists such as Andrew Lakoff (2005) and Kaushik Sunder Rajan (2006) have demonstrated that people in Argentina and India are now treated as human subjects for pharmaceutical drug trials for numerous reasons. Pharmaceutical corporations are finding it harder to locate a suitable population to test their new drugs in Europe and the United States because too many of the citizens are compromised as a result of the regime of drugs they are already taking. Argentinians are preferred as human subjects because they share many of the genetic similarities of potential European and American consumers. In India human subjects are preferred because there are fewer regulations and those selected are cheaper than subjects in Western nations. Another dimension of this new form of bioprospecting is that once the data is collected on these pharmaceutical human subjects in India, the data is sent to Europe or the United States in order to scrub the data and hide its origins and make it look as if the research is being conducted in the Western world by Western scientists. Such conduct works from two important assumptions. The first is that in order for science to be considered legitimate it still has to come from the Western world and second consumers of these drugs will easily accept any research results as scientifically valid as long as they seem to come from the Western world. In these cases the Argentinian and Indian subjects are constructed as an erasable people who invisibly serve the medical needs of citizens Europeans and the United States. These two examples I think put Paul Gilroy's (2000, 20) comments into a stronger focus when he writes "it is difficult to resist the conclusion that this biotechnological revolution demands a change in our understanding of 'race,' species, embodiment, and human specificity."

One of the major points about bioprospecting is that no matter if it is discussed in regards to colonialism or postcolonialism it is always a certain form of biopiracy in which the indigenous populations are less likely to benefit from the results. At the center of this reality is the ways in which indigenous populations have been erased as serious stakeholders when it comes to discussing the implications of bioprospecting. It has become easier for multinational corporations, empires, and nation- states to erase and ignore indigenous peoples than to respect their cultures, histories, and relationships with other peoples and the earth. No matter which word is used both imply that indigenous peoples do not matter unless they are a needed resource in extracting a plant, tree, mineral, animal, or human material. When prospecting is used to describe the history of botany or biology the word implies an innocence in which a person or corporate entity accidentally struck it rich by finding a use for a plant for instance that no one before knew about when of course indigenous people often did. Piracy, of course, implies something has been stolen; taken illegally but taken nonetheless therefore there is little the aggrieved can do to rectify the situation. Often race is a determining factor in defining the ways in which indigenous peoples interests are erased, this is what Gilroy would define as a form of raciology or the construction of a race to justify racism.

7.3   The Mythoi of and the Studying Up on Rice

Sylvia Wynter raises similar concerns as Gilroy's concerning the way in which the terms Indias and Negro/a were constructed during the Post-Columbian era. This construction of racial categories is part of Wynter's idea of mythoi and for this reason I wish to start with mythoi in this section regarding the story of rice in (post) colonialism.

Judith Carney and Richard Rosomoff (2009, 3) argue that "not all species transfers were mediated by naturalists and agents of trade. Ordinary people also instigated the geographical dispersion of plants and animals across the globe." African slaves were responsible for introducing rice, bananas, plantains, yams, and kola to the Western hemisphere. Yet until recently the historical record erased their role in these economic and cultural exchanges between geographical regions. In her own work on rice Carney demonstrates that this erasure was intentional and long lasting. "The historical botany of West Africa's chief food staple, rice, for instance," Carney (2001, 5) suggests, "was not widely known in the Anglophone world until the 1970s." It is finally in the 1970s that the historian Peter Wood began to note the origins of rice to the Western hemisphere as West African not Asian or Portuguese as the lore or myth suggested. Even then though Carney (2001, 5) notes "those contributions are still too frequently conceptualized as minor." The story of mythologizing the origins of rice in the Western Hemisphere is a story of Wynter's mythoi.

As the myth goes, the origins of rice in the Western Hemisphere is one of two possible places: Asian via Muslim trade routes into Africa or Portuguese sailors who brought this new technology of growing rice with them as they travelled into the Western hemisphere. Undergirding these myths concerning rice is the assumption that rice could not have been introduced by West Africans because the methods for growing rice successfully introduced to the Caribbean islands, South Carolina and Georgia were too sophisticated scientifically and technologically. This myth remains mostly intact in Western plantation lore. Carney (2001, 166) notes "the literature has yet to consider crops of African origin and the agency of Africans in their diffusion to the Americas. Africa remains conceptualized as a region where crops diffused to, rather than from, and where New World domesticates revolutionized the continent's agrarian systems." It is this erasure of history and the invention of an alternative explanation of the origins of rice that demonstrates Wynter's idea that humans are not just biological beings but also mythoic therefore hybrids. When Wynter refers to humans or any entity as mythoic or hybrids it is important to note she is not criticizing the human tendency to construct stories or myths to explain their existence. Wynter's embraces our poetic side because it is this poetic ability that fills in the silent gaps science creates. Instead, she is challenging the tendency of Europeans to construct the myth that humans are only biological. It is this myth of humans as biological beings that leads to the construction of the myth that West Africans could not have introduced rice to the Western hemisphere because, according to Man1 and Man2 European myths Africans are incapable of scientific and technological innovations as sophisticated as growing rice. Like Wynter, Carney wishes to tell a different mythoic (poetic) tale, but this one will look at historical records and evidence to construct a story of humans as both scientific and poetic beings.

Carney's reclaiming of the story of rice is an excellent example of Harding's notion of studying up because it literally begins with women and households and then builds from there to challenge structures of power and authority to include disenfranchised peoples into the story of scientific and technological developments. When the Portuguese arrive along the shores of West Africa what they discovered was a land of abundance, not a land of subsistence later Western myth makers claimed it was, and rice was at the center of this fertile land. The primary farmer of rice were women and depending on what area of West Africa they lived in two different methods were utilized to grow the rice. The first system where the tide waters effected the fertile land West Africans used according to the first-hand account of Andrè Alvares de Almada to grow "'their crops on the riverain deposits, and by a system of dikes had harnessed the tides to their own advantage.' This was the same system," Carney (2001, 18) notes, "European scholarship would subsequently attribute to Portuguese introduction." The second method was found further inland and referred to as a mangrove system. To describe this system Carney refers to historical record. This time it is eyewitness accounts of the slave ship captain Samuel Gamble who described the mangrove system as growing rice "'in quite a Different manner to any of the Nations on the Windward Coast… The country they inhabit is chiefly loam and swampy. The rice they first sew [sic] on their dunghills and rising spots about their towns; when 8 or 10 Inches high [they] transplant it into Lugars [places/fields] made for that purpose which are flat low swamps, at one side…they have a reservoir that they can let in what water they please, [on the] other side…is a drain out so they can let off what they please" (Carney 2001, 19). The amount of water that is placed in or released out of the Lugar depends on if it is the rainy season or not. During the fallow season the local cattle were allowed to roam the fields and naturally fertilize the area, creating a year-long cycle of land cultivation. While the women grew the rice, the men maintained the fields and levee systems. This is a system of labor that would continue once West Africans were forced into slavery in the Caribbean, South Carolina, and Georgia. It will also be the women who mill the rice via a mortar and pestle system.

Both approaches to growing rice were sophisticated methods. Carney notes that it was these methods that early slave trading Europeans would have observed while searching for slaves. These modes of rice cultivation proved to be dangerous and deadly for peoples of coastal West Africa. "Farmers, particularly those along the coast who planted irrigated rice," Carney (2001, 29) suggests, "were especially vulnerable: their sedentary way of life and proximity to European navigational routes made them easy prey to slavers. Capture or flight contributed to the breakdown of these more labor-intensive rice systems." As a result of this breakdown the once West African land of abundance became less developed and less able to supply the food needs of the population which eventually lead Europeans to conclude that Africa was not a land of abundance but of subsistence. As these methods of growing rice disappeared from West Africa, and transferred to the Western hemisphere, Carney (2001, 29) notes slave traders and European explorers moved further eastward into Africa and found "rice systems that required less infrastructure and labor," that is less of an irrigation system to maintain water level and less maintenance of the mangrove system" Without knowledge of the earlier sophisticated mangrove methods of growing rice the myth developed that Africans' forms of growing rice were not sophisticated and surely could not have contributed to the systems implemented in the Western hemisphere.

Carney also presents genetic evidence to reject claims that Asia is the source of African rice. If Asia were the source of African rice then the genetics of the rice would be the same or at least have some similarities. Carney (2001, 33) reveals French botanists in the late nineteenth century "encountered a type of rice distinctly different from the Asian varieties familiar to them." What the botanists eventually discovered is they identified two distinct forms of rice. The first was "*Oryza sativa* (Asian rice)" and the other was "distinct from Asian sativa...named *Oryza glaberrima*" whose origins were located in "the inland delta of the Niger River in Mali"(Carney 2001, p. 33–34).

By the time these West African methods of growing rice were transferred to the Western hemisphere another layer of mythologizing emerged. Early European observers noted that women cultivated and milled the rice and the men maintained the irrigation systems and fields. Because of this South Carolina and Georgia were unique in that proportionally and comparatively to other destinations for slaves women constituted a larger number of the slaves. At first, the knowledge that both female and male slaves possessed concerning the growing of rice provided them a little leverage to negotiate with the plantation owners. This allowed them to negotiate more time off to grow their own crops so they could survive the harshness of the "new" world. While most slaves were required to grow their own food few were given much productive time even on Sundays to do so. Rice plantation slaves were originally granted this "free" time. Slaves were able to negotiate other rights as well as Clarence Ver Steeg notes they "were not considered incompetent and semihuman; they were, in fact, entrusted with considerable responsibility, even to the extent of bearing arms for the colony's defense (Carney 2001, 100). As South Carolina and Georgia planters began to learn more about the West African system of growing rice, West African slaves were granted fewer human entitlements and privileges. Eventually what emerged is a myth that rice was developed in the Western hemisphere as a result of Asian trade routes and Portuguese technological ingenuity. After the technological transfer of rice cultivation was complete Carney (2001, 98) concludes: "No antebellum planters or, for that matter, their descendants considered the possible role of Africans in colonial rice development. The Africans were property and, as such, incapable of contributing valuable knowledge, skills, and experience." The institutionalization of Man1 and Man2 were complete.

In an attempt to challenge the myth of Asian and Portuguese origins of Western Hemisphere rice, Carney looks at the historical and scientific record. In her research, Carney cannot find any record that suggests Asian and Muslim trade routes came close to the rice growing areas of West Africa. "If Muslims had introduced the crop from Asia," Carney (2001, 35) deduces, "a geographic link to East Africa or the Middle East should be evident. None has yet to be discovered." Furthermore, Islamic historical documents demonstrate that when Muslims began to migrate into Africa rice was already being cultivated (Carney 2001, 35). In regards to the Portuguese introducing the technological innovations to Africa, I already mentioned that Carney cited European observers who chronicled how Africans already used two methods of growing rice that would be transferred to the Western hemisphere. Carney (2001, 49) further posits that when the Europeans tried to introduce different methods of rice cultivation to West Africa it was a failure and "the blame was always placed upon the presumed inability of Africans to comprehend the sophisticated concepts embodied in technology transfer rather than on the ill-conceived plans of outsiders."

By challenging the myth of European technological and scientific superiority and ingenuity and studying up from women rice growers and their households, Carney is able to construct a view of human relations that is simultaneously about the biology and mythoi of human beings. She is able to create a story of one crop that is both scientific and poetic thereby embodying Wynter's alternative to Man1 and Man2, the hybrid human who is both Bios and Mythoi; scientific and poetic. By using science and history together, Carney is able to challenge the myth that humans are only biological creatures. Moreover, by focusing on the role women played in the development of rice, technologically and agriculturally she offers a version of Harding's concept of science from below. She presents a convincing case why science should be explored by both postcolonial and curriculum scholars.

7.4 On Atlantocentrism and Postcolonial Thought

In doing the reading and research for these last two chapters what emerged for me was a tendency for postcolonial thinking to focus on those populations that were impacted by relationships with Europeans along the Atlantic coast lines. Perhaps it was just a result on the scholars I focused on. If this is the case then I am merely just continuing a tradition that constructs the Atlantic Ocean as the center of postcolonial debate. Nonetheless I think there is a distinct challenge present for postcolonial scholars to address. This challenge began for me after I read Jodi Byrd's (2011) *The Transit of Empire: Indigenous Critiques of Colonialism*. Although the works I focused on such as Norton (2008) who deals with indigenous cultures in contemporary Mexico and Central America, Bleichmar (2012) who deals with indigenous cultures in South and Central America, Carney (2001) who deals with African indigenous cultures, and Schiebinger (2004) who deals with Amerindians but mostly from the Caribbean, cover some indigenous peoples, North American indigenous peoples are often ignored. For Byrd (2011, p. xxxiii) it is important to recognize that there is such a thing as "'postcolonizing settler culture,' whose existence leads to "dispossessing indigenous peoples of home, land, and sovereignty." It is important to ask how has the locality of North American indigenous peoples been ignored and erased, how has bioprospecting in (post) colonial times done just what Bryd says: disposed indigenous peoples of home, land and sovereignty? It is important to ask how science and curriculum studies can address a less Atlantocentric perspective and develop approaches that include the complex questions and challenges indigenous scholars raise in light of, and sometimes in spite of, postcolonial thought.

References

Bleichmar, D. (2012). *Visible empire: Botanical expeditions & visual culture in the Hispanic enlightenment*. Chicago: University of Chicago Press.

Byrd, J. (2011). *The transit of empire: Indigenous critique of colonialism*. Minneapolis: University of Minnesota Press.

Carney, J. (2001). *Black rice: The African origins of rice cultivation in the Americas*. Cambridge, MA: Harvard University Press.

Carney, J., & Rosomoff, R. (2009). *In the shadow of slavery: Africa's botanical legacy in the Atlantic world*. Berkeley: University of California Press.

Chakrabarty, D. (2002). *Habitations of modernity: Essays in the wake of subaltern studies*. Chicago: University of Chicago Press.

Gilroy, P. (2000). *Against race: Imagining political culture beyond the color line*. Cambridge, MA: The Belknap Press of Harvard University Press.

Lakoff, A. (2005). *Pharmaceutical reason: Knowledge and value in global psychiatry*. Cambridge: Cambridge University Press.

Merson, J. (2000). Bio-prospecting or bio-piracy: Intellectual property rights and biodiversity in a colonial and postcolonial context. *Osiris, 15*, 282–296.

Norton, M. (2008). *Sacred gifts, profane pleasures: A history of tobacco and chocolate in the Atlantic world*. Ithaca: Cornell University Press.

Proctor, R. (2008). A missing term to describe the cultural production of ignorance (and its study). In R. Proctor & L. Schiebinger (Eds.), *Agnotology: The making & unmaking of ignorance* (pp. 1–33). Stanford: Stanford University Press.

Rajan, K. (2006). *Biocapital: The constitution of postgenomic life*. Durham: Duke University Press.

Schiebinger, L. (2004). *Plants and empire: Colonial bioprospecting in the Atlantic world*. Cambridge, MA: Harvard University Press.

Part IV
Interlude Four: Getting Lost and Finding Mary

Diana Brandi was a fellow graduate student in comparative education in the early 1990s at the University of Pittsburgh. She was passionate about bringing more diverse perspectives to the field. She was not the first feminist to enter the field. There were people like Gail P. Kelly who began to challenge the field before Diana came onto the scene, but Diana did write a review of *Emergent Issues in Education: Comparative Perspectives* (1992) edited by Robert Arnove, Philip Altbach, and Gail P. Kelly. What she challenged was the dominance of two perspectives within the field: Marxism and what she called functionalism and I would call neo-liberalism. Of course she received push back from the entrenched forces including disdain from some professors because she was a mere doctoral student. Who is she to question the authority of established scholars? I shared an office with Diana during this dispute. One day I picked up our phone. It was Patti Lather calling to voice her support for Diana. Here was an established scholar not accepting the orthodoxy but challenging the idea that two traditions could control and define the meaning of what is science, scientific, or scholarly. Patti Lather's stance for methodological pluralism continues to this day, and every field in education is a much better intellectual endeavor because of her work.

In her latest work, *Getting Lost: Feminist efforts towards a Double(d) Science*, Lather (2007) continues her work in rethinking educational research and academic work. Lather encourages us to adopt a Derredian approach to research. Lather (2007, 14) refers to as a "feminist double(d) science," that "means both/and science and not-science, working within/against the dominant, contesting borders, tracing complicity." What might it mean to be both doing science and not doing science simultaneously? For one, it means not making such arrogant assumptions that all work in the name of science is done as a form of universal truths when it fact it is merely a script, routine, and an ideology. It also means rejecting a methodocentrism in which one's faithfulness to an established methodological protocol is accepted as the only path to truth. It also means not accepting any binary that challenges the intellectual credentials of any scholar who challenges the orthodoxy of scientism or methodocentrism so prevalent in educational fields. To do science is to create one's own protocol, question grand assumptions such as the blind acceptance of the need for generalizability. To not do science is to place in doubt not the "subjects" who are objectified by researchers or the scholar who questions orthodoxy, but it is to encourage "getting lost" in a problem defined by oneself and others involved. "The task," according to Lather (2007, 10), "becomes to throw ourselves against the stubborn materiality of others, willing to risk loss, relishing the power of others to constrain our interpretive 'will to power,' saving us from narcissism and its melancholy through the very otherness that cannot be exhausted by us, the others that always exceeds us." To not do science is to recognize our work is important but it can never do everything that we think science demands we do. If science is science then scientists and scholars are the not science of this double. We are lost in our search for possible answers; knowing full well anything we find will be incomplete and any attempt to generalize or set in stone some protocol is merely a power play, a will to truth, not a will to power. I, for one, am content to be lost, only a cynic would exploit such lostness of another person. This is exactly what some will do in the name of "truth," "science," and "method."

What I think troubles those interested in science as a safety net and become skittish when they are lost, is the insecurity one has to admit when one does not know the answers to a problem or does not know in advance the way to best approach a problem. This is why when so many scientists discover a way to solve a problem they are quick to re-write history and pretend they knew all along what they were doing. This is what James Watson did in his memoir and rewriting the role of "Rosy" Franklin in this epic discovery. Watson and Crick stole her work and then tried to rewrite history by making her a complicit partner in order to cover up their caper.

In my career of lostness, I have wandered about, thinking about issues that shape academic work and challenge my assumptions about life. I have tried to pick topics that will not get me grants or easy recognition for a university. I have sought out topics that interest me and that expand a field of knowledge in the name of democracy and equality. Being lost allows one to explore with a freedom of inquiry that all universities say is one of their hallmark pillars as an institution, but in reality is merely words in a mission statement that means nothing. Freedom of inquiry means something to me and my life as a scholar bares the markings of this freedom. Being lost allows one to explore other scholars. If I did not embrace this freedom of inquiry I would never have read the work of Bruno Latour, Bernard Stiegler, Jacques Derrida, Robert Harrison, and for this next chapter Sharon Traweek, Karan Barad, and Vinciana Despret. After all what do these philosophers, animal psychologists, and anthropologists have to do with curriculum studies and education? Most people outside of curriculum studies ignore public education and it certainly frustrates me because a living democracy depends on a well-educated populace. But just because these scholars ignore our work does it mean we should ignore them. It is during my wanderings in lostness that I found Mary.

I think the work of Mary Aswell Doll is some of the more important work in curriculum studies. I have had the honor of knowing Mary for over twenty years, but I have had an even greater privilege of knowing her ideas. She is a fellow wanderer in lostness. The most influential of Mary's work on me is easily *Like Letters in Running Waters: A Mythopoetics of Curriculum*. In the introduction to this work and later in her memoir she repeats the following statement: "I would argue that the problem in our culture is not illiteracy, but the literalism that make us ill. Texts everywhere are being literalized: copied, imitated, mimicked." (Aswell Doll 2000, xiii). One cannot look at any religious denomination or political debate in the USA and not know Mary is correct in her observation. I want to twist Mary's statement around a little and connect her point to curriculum studies and science matters. What happens to a field of knowledge when it disengages from debates around science matters? It literally makes us ill. When we do not understand the power plays of pharmaceuticals to "scrub" data that supports their new wonder drug or buries evidence that disputes the effectiveness of a new drug, then our disengagement will make us ill. When we do not understand the impact of climate change and just shout it is true and deniers are ideologues while not understanding the evidence or the process of data generation from models, such disengagement will not literally make us ill. It will erase us from the planet.

Another wonderful statement from my found lost mate Mary is: "When people say they are frustrated I think they lack art." (Aswell Doll 2000, xvii) Again I twist. When people ignore the art and importance of/in science this disengagement leads to frustration and a malaise threatening the vitality of a living democracy. Science, like any human endeavor, deserves the attention of curriculum scholars because it is an art, and to lack art is to lack hope.

In the next chapter I look at the work of Sharon Traweek, Karan Barad, and Vinciane Despret to explore what feminists scholars bring to our understanding of science. Their work is art and it is a cure to our illness.

References

Arnove, R., Altbach, P., & Kelly, G. (Eds.). (1992). *Emergent issues in education: Comparative perspectives*. Albany: SUNY Press.

Doll, A., & Mary. (2000). *Like letters in running water: A mythopoetics of curriculum*. Mahwah: Lawrence Erlbaum Associates.

Lather, P. (2007). *Getting lost: Feminist efforts toward a double(d) science*. Albany: SUNY Press.

Chapter 8
Of Hierarchies, Cultures of No Culture, Ontology, Protocols, and Anecdotes: (Re-writing) Women and Science

Yes, we were opportunist amateurs, shamelessly meddling in fields where our authority was not recognized, called by the feeling of possibility, by events that awaken the sense of adventure where dilemmas seem inescapable.
Isabelle Stengers and Vinciane Despret, *Women who make a fuss*, 69.

8.1 Traweek's Anthropology

When I receive a book in the mail or buy one at a bookstore I will do two things with it once I am home. I will hold it, look at its cover and feel and smell it because all books, new or older, have a certain touch and smell to them. I will also peruse its contents. When I am perusing I will decide in my head when I will read the book in relationship to other books and what contents of the book I will read. Some single authored books this is difficult to do so I dive right in and read it until I decide if it is a book for me or it is time to find another book to start. Edited books are different and easier to decide what will be read. This is how I began reading in feminism and science. In either 1992 or 1993 I went to my favorite bookstore in Pittsburgh, Borders (remember them?). I was just beginning to read in science studies and came across Andrew Pickering's edited book *Science as Practice and Culture* (1992). I was definitely going to read Pickering's introduction and I was intrigued by Karin Knorr Cetina's work and Michel Callon and Bruno Latour's piece I would read because I already read their other work and liked it, same for Steven Fuller's. I decided to also try out for the first time Henry Collins and Ian Hacking's chapters. They both since have been educating me with delight. The last chapter did not grab my attention. Ignoring the first part of the title, "Border Crossings: Narrative Strategies in Science Studies…" I focused on the last part "among Physicists in Tsukuba Science City, Japan." What? Who is interested in Japanese physicists except for maybe Japanese physicists and at least one anthropologist? I was not. After I read the rest of the book, I thought what the hell I might as well read the last chapter. Since that moment in the

© Springer International Publishing AG, part of Springer Nature 2018 165
J. A. Weaver, *Science, Democracy, and Curriculum Studies*, Critical Studies of Education 8, https://doi.org/10.1007/978-3-319-93840-0_8

early 1990s Sharon Traweek has been a part of my education in science studies and for obvious reasons. It is Traweek who demonstrated to me with clarity and persuasion the importance of observing gender formations and relations within any academic culture including the natural sciences, in shaping the meaning of knowledge and facts. In this concluding chapter Traweek (1992, 443) tells the story of sharing her work with a MIT women faculty meeting. After discussing the ways in which gender shapes the sciences a junior faculty member approached and proclaimed "that the issues we wanted to investigate simply did not exist in the scientific community". This faculty member was not alone as the other junior members seemed to agree with her. This, of course, is not the story of my reading into gender matters. After the junior faculty member spoke, a senior faculty member intervened and said "the junior faculty did not yet see gender as an issue in their work because they had not yet gotten to the career stage at which they would define fully independent research projects requiring their own command of significant resources." (Traweek 1992, 443) These words had a profound effect on me. At the time I was working on my dissertation dealing with the politics of academic work in the context of (re)unified Germany and the community of historians. Traweek provided a stark example of how knowledge formation was political and what counted as knowledge was shaped by the position one held within a hierarchy of scholars and how much influence one had in shaping research agendas. This was exactly what I was observing in my visits to Germany and reading in my interview transcripts. Traweek's influence on my thinking continues to this day.

There are two other ways Traweek's work influenced my thinking. While I was revising my dissertation into a book on comparative academic politics between German historians and American academics I wrote about the so-called "science wars." I do not use in my writings such hyperbolic words such as wars to describe academic debate because in times of endless wars on terror, drugs, immigrants, and other peoples I know of no one who died because of the intense and sometimes irrational debates over the importance of science studies to scientists. Careers were altered and smeared but no one was forced to immigrate because their homes were destroyed by "strategic bombings" and no one was innocuously labeled collateral damage and buried in an unmarked, silent grave. During these debates Traweek wrote a piece called "Unity, Dyads, Triads, Quads, and Complexity: Cultural Choreographies of Science." In this chapter Traweek (1996, 148) raised an important point when she asked "Why should there be only one way to think well, only one way to have fun with our minds? Why is mental monogamy required? Are we still fighting about monotheism, Manichaean fallacies, and Albigensian heresies?" This short phrase has shaped almost everything I have written since I read her piece in the late 1990s. Why is their supposedly only one way of thinking about important work, one way to "do research," and why is it this one way seems to be ahistorical as if it were presented to human beings at the beginning of time and it took the British philosophers Bacon, Boyle, and Newton and later the social scientists to perfect it and present it to all novices entering the fields of science and the social sciences as The Way? Her use of the word monotheism and monogamy were perfect because they highlight the religious undertones one felt when entering the most holy of holies, the laboratory.

Science did not have just have an aura of being ahistorical, untainted by human hands or minds. It was as Traweek (1988) noted in her book *Beamtimes and Lifetimes* "a culture of no culture." This was the third way Traweek shaped my thinking. In her anthropological study of particle physicists Traweek noted that this culture of no culture was a world in which these physicists wished to live or even worse pretended to live as if it really existed. This culture of no culture was "an extreme culture of objectivity…which longs passionately for a world without loose ends, without temperament, gender, nationalism, or other sources of disorder—for a world outside human space and time" (Traweek 1988, 162). Her book is a chronicle of how physicists tried to create this world in both the United States and Japan. It is also about how as an interloper she crossed these patrolled borders (not necessarily by physicists though) and looked at the culture that claims no culture in order to show there is culture in the sciences and it shapes everything scientists do. Her work is as important as Bruno Latour and Steve Woolgar's (1986) *Laboratory Life: The Construction of Scientific Facts* or Latour's (1987) *Science in Action* but she does not garner as much attention as these other scholars. In this chapter I want to focus on Traweek's work dealing with the hierarchies of knowledge, the education of scientists into the culture of no culture, and the role that gender plays in the formation of knowledge and (non)cultures. I also want to focus on the work of Karen Barad and Vinciane Despret to demonstrate how feminist science continues to grow and challenge the culture of no culture in order to break from the strangle hold of mental monogamy and create different ways of seeing "nature" and different ways of "doing science." In the field of educational research mental monogamy is still the norm. Being in a college of education is sort of like being in a Catholic Church that refuses to acknowledge that the Reformation took place 500 years ago. Yet as these three scholars demonstrate there are so many other ways to look at research and to think differently.

By the time Traweek's book *Beamtimes and Lifetimes: The World of High Energy Physicists* came out in 1988 she was already studying this scientific culture for 15 years. Apparently a short time in Anthroplogical time. In her ethnographic work she was interested in "what the community takes to be knowledge, sensible action, and mortality, as well as how its members account for troubling information, disturbing actions, and troubling motives" (Traweek 1988, 8). She was interested in how the actors in this community played the game of science. "Like many social groups that do not reproduce themselves biologically," Traweek (1988, 74) noted, "the experimental particle physics community renews itself by training novices…. This transmission of meaning occurs not only in formal education, but also in the daily routines and in 'the informal annotations of everyday experience called common sense." In both realms, formal education and the construction of common sense, it is a man's world to define. Nonetheless, it is a formal and informal education based on competition, hard work, luck, persistence, and networks. As Traweek (1988, 75) acknowledges, "about 75 percent leave the field after [their] fifteen-year training period." Time is not on the novice's side. It is a matter of getting on the right team, with the right research agenda but the senior researcher will not just pick any-one to be on his (it of course is mostly a he) team. The graduate student must show promise academically and culturally (intuitively). As Traweek (1988, 82–83) observes the graduate students "are learning to become meticulous, patient, and per-sistent, and that these emotional qualities are crucial for doing good physics. They also are beginning to learn what is meant by 'good taste,' 'good judgment,' and 'creative work' in physics….The novice is learning by a process of trial, error, and comparison." If this student is good enough, has assimilated the culture well enough, and demonstrated an ability, "the advisor is expected to make use of his network to find the student a postdoc position" (Traweek 1988, 82). There is an art to this pro-cess too. If the right post-doctorate position is found it also means a growing net-work for the senior physicist who now can rely on someone to make connections with another research group and maybe share results and information that can be fruitful for the senior member. However, if the student is not as good as the senior researcher thought then this reflects negatively on the researcher who might then gain the reputation of nurturing students who are not as stellar as they should be in understanding physics and the inner works of the laboratory and the physical realm. "It is widely understood among the senior group members in American labs," Traweek (1988, 87) notes, "that it is not sufficient for a postdoc to do" mundane work; "independent, risky work [research with the potential for rich results but only within the bounds of accepted physics] must be undertaken as well….Independent, risky work can be undertaken only if the postdoc succeeds in gaining sole responsi-bility for some project, a privilege not readily granted. This opportunity is won by displaying a convincing faith in one's own power to do a task better than others in the allotted time and within the budget, no matter what the obstacles." To get to this point of actually earning some independence to get some "beamtime" (or working with the accelerators or colliders) and creating some potentially productive data, gossip or talk is very important. "What is accomplished among the physicists by talking" Traweek (1988, 117) asks? "Primarily they are evaluating their peers and their work, persuading those same peers to support their work, managing the distri-bution of news, arranging for positions for novices." Through gossip or talk the sci-entists are trying to convince their peers that the data extracted from their beamtime are real data not erroneous and are worth their attention. The attention of a physicist is always a matter of priority because the field is changing rapidly. As Traweek (1988, 120–121) observed the rapid pace of developments in the field means that "Good experimentalists do, write, and talk physics, but they rarely read physics. Important results are usually written up quickly and are available for 'preprints' within a few weeks of discovery…preprints appear in journals within weeks…wait-ing to learn of interesting data…until they appear in the journals is regarded as exceedingly unwise. What is being talked about is the current, more advanced knowledge; what has been written is considered established, uncontested, and hence uninteresting." The gossip or talk "work" physicists do to establish data as fact worth publishing in a journal but more importantly establishing a line of research that is defined as promising or worth beamtime and budgetary allotments is according to Traweek (1988, 121) "judgmental" with "real sanctions and rewards." The results from this "talk" is where physicists earn their reputations. "High energy physicists need good reputations to stay in the community; the loss of a good name means isolation and even expulsion."

What emerges from this culture of high energy physics is a hierarchy of knowledge. This knowledge is not just about turning potentially erroneous data into meaningful data resulting in a fruitful research agenda and eventually the establishment of a fact. It also represents a worldview of knowledge and gender relations determining who can create paradigmatic worldviews. Traweek notes that particle physicists embraced a traditional hierarchical notion of knowledge formation that can be traced as far back as Kant who believed philosophers of physics and mathematics were at the pinnacle of knowledge with actual physicists, astronomers, and later chemist right behind them and biologists and social scientists some distance behind, and holding up the bottom were the literary critics, historians, artists and rogue anthropologists like Traweek. The particle physicists Traweek observed came to this hierarchical form of knowledge because people left particle physicists to enter other fields such as astrophysics and chemistry but no one leaving those fields entered particle physics. Social scientists and those in the humanities most definitely never entered the realm of particle physics. "Significantly," Traweek (1988, 79) notes "the fine arts and mathematics are not ranked in this hierarchy of knowledge. Each of them is thought to share with particle physics certain crucial characteristics: art and physics both require creative imagination...particle physics is presumed to include what is best about art and mathematics, while excluding the rest."

Where do women enter into this hierarchy of knowledge and what role does gender play in shaping the field of knowledge called particle physics? If women enter into the hierarchy it is as a physicist but this is rare. Traweek suggests about 3% of the physicists are women. Even when they appear in the dramatic performance called particle physics Traweek (1988, 16) remarks early in her book that during her 15 plus years of visiting various laboratories "the status of women within them has remained unchanged...women remain marginal," Traweek was not exempt from this marginalization either. During her ethnographic work she reports being mentored by a professor at her first academic position at MIT who noted "after five years or so [she] could move on to a university which had a visible anthropology department, any one of which would also be impressed because [she] had been validated by scientists and engineers." She continues: "Although you are perhaps reading an exception, to my knowledge no institution and no press has seen fit to hire me, promote me, or publish my work without first getting the opinion of extremely prominent physicists" (Traweek 1992, 441). Traweek's decades of work was not validated or sanctioned as a form of knowledge, apparently, until a physicist stamped his approval on it. It was an authority, Traweek notes that physicists did not insist upon. "I hasten to remind you, Dear Reader, that the physicists did not ask for this authority over my stories; it has been given to them by my senior colleagues in anthropology and science studies and by university presses." Stereotypes in spite of shifts in thinking die hard.

Times have changed though and Traweek's work from 1988 is obviously dated. Physics of any kind, with the possible exception of astrophysics, has lost its dominant position of knowledge production to biology while many have questioned the whole reality of a hierarchy of knowledge of any kind let alone one where particle physics remains at the apex. Yet, it seems her premise remains salient in spite of many changes occurring in society and the sciences. In her recent work at UCLA, Traweek has been working with information studies faculty, historians of science, women's studies, education faculty, and other anthropology colleagues to look at the various issues centered around the creation of digital libraries for data generated from big and small science projects in astronomy and computer science. This group is interested in "what new infrastructures, divisions of labor, knowledge, and expertise are required for data-driven science?" and "What data are most important to curate, from whose perspective, and who decides?" (Darch et al. 2015, 63). For Traweek her institution seems to have changed from Rice to UCLA, her field of interest seems to have shifted from particle physics to astronomy and computer science and she appears to be working with a research team rather than observing research teams. What has not changed, however, is her interest in gender relations in the sciences. In their work looking at astronomers and data, they have noted that a new hierarchy of knowledge is emerging as the science fields change. "In our interviews" Murillo et al. (2013, 38) note "the demographics of the emerging subfield [of astroinformatics] are composed mostly of under-represented groups [women and non-Europeans]. Some but not all carry the title of astronomer. In contrast, fewer established European American men astronomers are as engaged with astroinformatics; they secure the funds for the informatics portions of their projects, and hire others, sometimes called technical staff, to do the work." In other words, women still find themselves in a subordinate position to define what knowledge is worth most and who shall conduct the research and collect the data. Once the grants are written and awarded, the research conducted, and data collected then it still appears women's work just begins; the work the men seem very disinterested in doing.

8.2   Karen Barad and a New Ontology of Science

Meeting the Universe Halfway: quantum physics and the entanglement of matter and meaning I believe is the most influential book written in science studies since Kuhn's *The Structure of Scientific Revolutions*. Barad's book requires readers to accept the importance of Bohr's philosophy and thinking about what nature is and how scientists work. As a result of Bohr's work Barad pushes us to rethink our fundamental epistemological and ontological conditions. Recognizing Bohr's centrality in thinking about science makes this book important by itself but where her work is most influential is the directions she pushes Bohr's ideas beyond anything he might have been comfortable entertaining. Getting someone or some concept to do more than they are prepared to do is a mark I believe of an important contribution to a field of knowledge. Just as Derrida pushed Rousseau to reveal more than perhaps he might have wished or Joyce challenged accepted notions of the novel, Barad is pressing the boundaries of science and key concepts that shape it. It is a risk when a scholar pushes accepted boundaries. She not only risks the acceptance of her ideas but also her reputation within a field of study. I think Barad's risk worked. In this section I wish to demonstrate why Barad's ideas are important for curriculum studies scholars and how they require us to rethink traditional notions of science and research such as objectivity, reality, and nature.

To begin with Bohr and light. During the debate over waves and particles, Bohr entered into the discussion and introduced his important concept of complementarity. "For Bohr," (Barad 2007, 106), "the crucial point is the fact that wave and particle behaviors are exhibited under *complementary* (all italicized words in the quotes in this section are Barad's)—that is, *mutually exclusive*—circumstances. According to Bohr, either we can find out which slit an electron goes through by using the which-path apparatus,…or we can forgo knowledge about which path the electron goes through (using the original unmodified two-slit apparatus) and obtain a wave pattern—we can't have it both ways at once." Implicated in this passage is any notion that suggests a measuring device is separate from that which is being measured or a human creation is separate from nature. Barad (2007, 106) continues: "this all seems very sensible, but the implications are nothing short of revolutionary. Notice what the complementary nature of these results mean: *the nature of the observed phenomenon changes with corresponding changes in the apparatus.*" An apparatus (such as Bohr's two slit diffraction device), a statistical approach, an observation in an ethnographic study, or an anecdote will shape how nature or reality is viewed, interpreted, and shaped. If there is a modification in the approach it will change reality. With his experiment

> Bohr called into question two fundamental assumptions that support the notion of measurement transparency in Newtonian physics: (1) that the world is composed of individual objects with individually determinate boundaries and properties that can be represented by abstract universal concepts that have determinate meanings independent of the specifics of the experimental practice; and (2) that measurements involve continuous determinable interactions such that the values of the properties obtained can be properly assigned to the premeasurement properties of objects as separate from the agencies of observation. (Barad 2007, 107)

In other words, Bohr's notion of complementarity places in doubt the ideas that human constructs created to probe and discover nature are separate from an independent reality and what is obtained through an experiment corresponds to what is out "there" in reality. For Bohr, "*there is no unambiguous way to differentiate between the 'object' and the 'agencies of observation.' No inherent/Cartesian subject-object distinction exists…the measurement interaction can be accounted for only if the measuring device is itself treated as an object, defying its purpose as a measuring instrument*" (Barad 2007, 114). As an object the instrument becomes a part of a reality creating relationship with that which is being measured. The object and instrument become one. What is captured then is not a reality out there but a relationship between the researcher, an instrument, and another object. As a result, "referentiality must be reconceptualized. The referent is not an observation-independent object but a phenomenon. This shift in referentiality is a condition for the possibility of objective knowledge. That is, *a condition for objective knowledge is that the referent is a phenomenon* (and not an observation-independent object)" (Barad 2007, 120). After Bohr the world is no longer a clean, orderly Newtonian or Cartesian realm in which subjects can be separated from objects, each object has its own reality, observers can distance themselves from the observed, and observing devices do not interfere with the reality it reveals. "Bohr's philosophy-physics undermines a host of Enlightenment notions, requiring him to construct a new logical framework, including a new epistemology, for understanding science.... Measurement practices are an ineliminable part of the results obtained. Since these practices play a crucial role in the world, they must be a part of scientific theorizing; that is, Bohr situates practice within theory" (Barad 2007, 121). Bohr does not suggest that objectivity is now impossible; he just changes the notion of objectivity. A traditional, nineteenth century notion of objectivity, that is to say a mechanical or structural notion of objectivity invented to disconnect the subjective observer from the object observed, is now untenable, although still practiced by too many in everyday life, journalism, politics, educational research and other realms of life (Daston and Galison 2007). Objectivity now becomes an entanglement between phenomena and the communication of these intra-actions. "Since individually determinate entities do not exist," Barad (2007, 128) notes, "measurements do not entail an interaction between separate entities; rather, determinate entities emerge from their intra-actions. I introduce the term 'intra-action' in recognition of their ontological inseparability, in contrast to the usual 'interaction,'...*a phenomenon is a specific intra-action of an 'object' and the 'measuring agencies'*". In other words, "objectivity for Bohr is not a matter of being at a remove from what one is studying, a condition predicated on classical physics' metaphysical belief in individualism, but a question of the unambiguous communication of the results of reproducible experiments" (Barad 2007, 174).

Although the consequences of Bohr's principle of complementarity are profound, Barad (2007, 337) is clear to note that "Bohr was not in the business of doing philosophy for philosophy's sake; for Bohr, the physics guided and motivated the philosophy, not the other way around. As a physicist, Bohr's primary concern was the understanding of experimental outcomes." If anything is extrapolated from Bohr's philosophy beyond implications for physics, Barad is going to have to venture into an unknown realm and do it herself. This is what makes her work an important book. Barad creates a new philosophy that impacts not only epistemological issues for physics but more importantly she creates a philosophy that raises radical ontological issues facing scholars in a post-human world of the twenty-first century and beyond. For Barad, Bohr's philosophy is steeped in an anthropocentric mind-set in which he "insists that 'humans' be understood as 'parts of nature,' and in a second breath, he privileges 'humans' as special envoys sent out to secure the grounds for objectivity" (Barad 2007, 330). Humans become the master communicators of all of nature and the purpose for any experiment in all intra-actions between devices and objects. Barad wishes to challenge the humanism embedded in Bohr's physics-philosophy. She proposes a posthumanist performative approach that "specifically acknowledges and takes account of matter's dynamism. The move toward performative alternatives to representationalism shifts the focus from questions of correspondence between descriptions and reality...to matters of practices, doings, and actions....Posthumanism, as I intend it here, is not calibrated to the human; on the contrary, it is about taking issue with human exceptionalism while being accountable for the role we play in the differential constitution and differential positioning of the human among other creatures (both living and nonliving) (Barad 2007, 135–136). The human remains in the intra-actions of the universe but is not central. Humans play an important ethical role as Barad mentions. To constitute and position various phenomenon (both living and nonliving) into an entanglement or a system of ordering requires humans to account for their doings, to think about the implications of these doings not just for humans and their notions of reality or objectivity but for everyone and thing involved in the new entanglements. Despret's work in the next section will show exactly what these implications for and consequences of a posthumanist performativity are. In a posthumanist performative ontological system humans are not arbiters of what reality is and what is valued. They are one part of a system; a reality of phenomena shaped by the apparatuses created, the problems defined, the meaning construed from the phenomena, and the ethical consequences gleaned from the intra-actions.

It is important to note here that Barad is not reinstituting a new non-humanist form of objectivity in which the human is banished from "nature." As already stated she recognizes Bohr's humanist assumptions but "it is important, however, to distinguish between a principled exclusion of the human, based on the belief that humans have no place in a physical theory, and one based on the posthumanist refusal to presume that humans occupy a privileged position in physical theories." (Barad 2007, 323) One approach privileges the human as a god, all seeing and all knowing; able to determine what is truth, nature, and reality while at the same time wishing away all traces of humans from creating human interpretations and apparatuses to understand reality beyond humans. The other approach makes room for non-humans to emerge in reality and in the laboratory while recognizing the limits of human creation and thinking. This making room for non-humans (living and non-living) is Barad's beginning move to extend the implications of Bohr's physics-philosophy. It is Barad's first ethical move. Non-humans are no longer objects existing in the universe waiting for humans to discover them and provide meaning for them. Non-humans are already moving into and out of intra-actions, shaping realities and creating meanings. It is not their responsibility or purpose in the universe to demonstrate their utility to humans, nor is it to submit to the will of humans to construct them as utility. Their role is to be. As Barad (2007, 338) suggests "human concepts or experimental practices are not foundational to the nature of phenomenon. Phenomena are not the result of an external imposition of human-based conceptual schemata. Rather, phenomena are the manifestation of material-discursive practices, where discursive practices are not placeholders for human concepts but specific material articulations of the world." What these material and discursive conditions are is not a predetermined notion or even human determined. "Phenomena are not mere human contrivances," Barad (2007, 338) insists, "manufactured in laboratories. Phenomena are constitutive of reality…. There is no preexisting, separately determinate entities called 'humans' that are either detached spectators or necessary components of all intra-actions." Like all other entities humans are not the referees, officiating and dictating the meaning and flow of reality from some secret, even to themselves, Archimedean point. They are intra-acting players part of the universe sometimes part of a reality, other times, most of the time, never part of a reality. Once this limited role in reality is accepted the most important aspect is to ask what is the impact of any human intra-actions and what is the price for entering into new intra-actions with nonhumans? Traditionally, in science, let alone any dimension of life, this type of question is never broached and if it is it appears only later when humans have entered an intra-action while rationalizing why humans are simultaneously central to and self-exiled, as an objective observer, from any intra-action. In Barad's posthumanist performative universe no entity can claim a neutral and universal role in any intra-action. Traditional objectivity under Barad's intra-action universe is untenable. It is merely an illusionary step to create an alternative universe that can never exist in reality. Why then do so many in educational research and so many other realms of scholarship insist in living in this illusionary world? What are the ethical implications of such illusionary moves?

8.3   Despret's Anecdotes and Other Non-human Animals

I want to start this section with ravens. Why do ravens usually live solitary lives or as "couples" but they will usually not eat alone but in a large group? This was one of the problems Bernd Heinrich faced when, against the advice of his mentor, he decided to study Maine ravens. Why would a solitary animal such as a raven discover a carcass, wait, watching the dead animal, for as long as 3 days to take its first bite, and then alert other ravens so they would come to feast or watch others partake? Could they eat in a group because they are leery of the so-called carcass who may be playing opossum in order to pounce on the first raven that tries to take a bite or is the actual carcass a trap set by other predators waiting for the ravens to take the bait? Could it be just as traditional scientists conjectured that like other animals ravens are merely putting evolution into action and while one raven takes a risk of ignoring the potential threats in the environment the others serve as look-outs? The ravens would then just be another example of science in action, serving its inner need to eat while perpetuating, just as evolution predicted, the species while those who missed this gene line would surely act too hastily, not wait for back-up, and pounce on an assumed meal that could be playing dead or embody the site of an ambush. Heinrich was not convinced that what humans had devised as raven behavior was accurate at all. While studying ravens Heinrich took in an injured raven and this raven did not act the same way ravens did in the wild. As the theory of non-agential animals goes, the raven would just follow instinct, not trust Heinrich to feed it until the raven was convinced he was not a threat, and then the raven would finally eat. Yet, under the circumstances the raven was quick to take Heinrich's offerings. After years of observations and listening to what ravens had to tell him, Heinrich found out that the group feeding mentality was not as evolutionary, non-agential as scientists thought. Ravens who summoned other ravens to partake or witness a feeding were not doing it for any altruistic, gene propagating reasons, they were actually interlopers who were entering a raven couples' territory and eating their food. The stranger raven was raiding the refrigerator of a raven couple and needed allies to witness the raid so the couple would not chase off the interloper. Raven couples will defend their territory when only a single or a few ravens appear, but when a mass of ravens are in sight the couple allows interlopers to feast. The interloper calls upon other ravens to act as insurance when it is entering the territory of an established raven couple. Instead of trying to control the environment so he could report something to other scientists who would confirm onto Heinrich the title of originator of something reliable and factual therefore original, Heinrich accepted that with ravens it "'depends more on how well one follows the situation than on how well one controls it'" (Despret 2015, 67). By following the ravens, Heinrich discovered how resourceful and adaptive ravens are. "The primary achievement of a raven," Heinrich explains, "is first and foremost that it 'can procure resources from the environment and convert them to more of itself'" (Despret 2015, 61). In Heinrich's work a raven went from being a less than intelligent non-human animal with the power of instinct and the evolutionary desire to preserve the gene pool to an opportunistic, inventive, adaptive non-human animal with agency to enter in and out of various environments and communities.

I start this section with this specific anecdote because as a reader I found it to be the most interesting of anecdotes Despret offers in her collection of work to demonstrate how certain scientists are trying to think differently about other non-human animals. While Traweek's contribution to science studies is how gender is shaped in science and how individuals are educated to think in specific (gendered) ways in terms of what is knowledge and Barad rethinks objectivity and intra-actions between humans, and other living and nonliving entities, Despret's many contributions to science studies is to encourage scientists to see the limits of their protocols, their models (what she refers to as dispositives), and how these limits shape what is accepted as reliable knowledge. The use of anecdotes is one way she undermines the dominate way of thinking in science. "Anecdotes are coming back with greater frequency," Despret (2016a, b, 75) asserts, "and a renewed vigor in essays today that plead in favor of the better treatment of animals, even their liberation. Animals that run away, revolt, or are aggressive toward human acts deliberately, as attested by their rebellion of conscience to the injustice of which they are victims." Anecdotes become a means to challenge set protocols and methods scientists use to define and control other non-human animals. With the rise of ethology as a science, one way scientists marked their territory of reliable knowledge was to degrade anecdotes as unreliable and mere hearsay. Despret reverses this cultural trend and uses anecdotes to demonstrate numerous aspects about animal life that humans miss when they rely more on protocols and set notions of what an animals is and can do. Anecdotes become a means to restructure science.

As part of this demarcation process between those who are scientific and those who are not, Despret notes that the knowledge of rivals, "amateurs. breeders, and trainers," had to be dismissed as "anecdotes and hopeless anthropomorphisms" (Despret 2016a, b, 34). As a result of this process of demarcation early creators of ethnology like Konrad Lorenz "will 'scientize' the knowledge of animals. Ethology becomes a 'biology' of behavior, hence the importance of instincts, invariant determinisms, and innate physiological mechanisms that are explicable in terms of causes." Protocols and methods would be set in order to control the flow of knowledge to convert animals' actions into reliable knowledge, replicable and generalizable, the true hallmarks of a real science. The folksy wisdom of herders and breeders can remain in the pastures, stables, fairs, and shows while the scientists do the important work in their laboratories. What emerges for Despret (2016a, b, 40) from this demarcation between science and the amateurs is a "strategy of 'doing science,' as a procedure of placing at a distance those who might claim to know (and how to know),...translat[ing] itself into a series of rules. Thus the rejection of anecdotes (that so remarkably punctuate the discourse of amateurs) and above all the manic suspicion with regard to anthropocentrism appear as the mark of a true science. Scientists who inherit this history now manifest an intense distrust of any attribution of motive to animals." With the distrust of anecdotes and the distrust of anthropocentrism rises a science, ironically steeped in anthropocentrism. In an attempt to limit the construction of non-human animals with human characteristics, scientists created a protocol of human behavior based on human assumptions of how non-human animals act. Under scientific anthropomorphism and in the name of anti-anthropomorphism non-human animals became instinctual, unemotional, non-agential, and different from superior humans. Scientific anthropocentrism created protocols to glean information from mice in mazes and other animals in other controlled settings, but this turned non-human animals into anthropomorphized research subjects, limited in abilities by human assumptions and a lack of human imagination and inquisitiveness. The system known as ethology did create scientific knowledge but at a price. It turned non-human animals into hierarchical beings and humans into administrators of often cruel protocols.

Despret shares the story of Harry Harlow. Harlow was interested in rats and wanted to study the "choices made by unweened baby rats: would they prefer cow's milk to other liquids? Would they accept orange juice in the absence of material milk?...Harlow remarks that the baby rats stop eating when the air is either too cold or too hot. Only a temperature equivalent to that of a mother's body seemed to favor intake...From there it was a small step to the idea that the babies might perhaps prefer to be with their mother" (Despret 2016a, b, 147). To test his idea Harlow devised ways to separate mother from babies including cages with barriers and mother starved and separated from babies. The mother always found a way to get to the babies. Is this behavior "a reflex...An instinct?" (Despret 2016a, b, 147). To determine this Harlow "removes the mothers' ovaries, blinds them, detaches their olfactory bulbs. Blind, without hormones, and even without smell, the mothers continue to rush straight toward their babies." This drive for Harlow was "the need for contact" (Despret 2016a, b, 147). For Despret the issue is not the need for contact among rats but what drives a scientist to such "rationalized" behavior to mutilate another living being to try to prove non-human animals have similar human drives, needs, desires? Despret (2016a, b, 149) notes "the true thread that guides this story appears: that of a routine that loses control and becomes mad. Separating mothers and their babies, then separating mothers from themselves, from their own bodies, removing their ovaries, their eyes, their olfactory bulbs—what is known as the model of 'breaking down' in science—separating for hygienic reasons, then just for separation itself." What happens when scientific protocol is out of control, turning non-human animals into objects for torture and exploitation? Through the use of anecdotes Despret offers up a stinging critique of scientific knowledge acquisition while she searches for a different way to do science; to see the world, to be in the world with other animals.

"There are two different ways of doing science...in the area of animal studies" Despret (2016a, b, 114) suggests..."On one hand we have a method inherited from biology and zoology that looks for similarities and invariants, within species...by requiring animals to obey [human] laws that are susceptible to generalizations and to relatively univocal causes that can be inscribed within an interpretive routine...On the other hand, we have another practice that is in competition with the former; this one is inherited...from anthropology's ways of thinking and doing and seeks to explore by focusing on their flexibility, the singular and concrete situations encountered by animals." These two ways of seeing are "about politics, but the politics of science and of political relations to nonhumans." While I think Despret is limiting the possibilities of doing science to a modernist binary between biology and anthropology, her point is clear. What is it humans are trying to get out of animals? Are humans interested in fitting non-human animals into their worlds, their laws, their protocols or are humans interested in a different relationship with non-human animals? Clearly, Despret is interested in the latter and demonstrates how it can be done, as I am since this is why I include her in this chapter. Despret offers a different way of seeing the world and thinking differently within, around, outside of science, sometimes in order to rethink science and other times in spite of what is accepted at the moment as science. In her work, Despret offers numerous ways to think differently about non-human animals and science. One interesting point she stresses is a need to accept what she calls a more "anthropo-zoo-genetic practice."

Of course, to describe what this neologism means she provides two anecdotes. The first one is Hans the horse. Was Hans the horse a mathematical genius? Could Hans think mathematically when approached with a problem? Apparently Hans could until Oskar Pfungst arrived on the scene and noted that Hans was not acting on his own but reacting to human cues that the humans did not know they were sharing with Hans. When Hans was given a mathematical problem, he would use his hoof to relay the answer. What Pfungst discovered was when Hans came to the right number the humans around him moved their bodies in anticipation of the impending right answer and Hans at that moment would stop thereby getting the right answer. "How could it happen", Despret (2004, 116) asks, "that humans replace their own spontaneous movements with that of the horse, unless we assume that Hans taught them the gestures he needed?" Hans was as much the teacher as were those who were giving him the problems. Humans inter(intra-)acted with Hans to create meaning. They were partaking in an "anthropo-zoo-genetic practice."

In the second anecdote a psychologist named Rosenthal gives his students a rat. Some of the rats he tells them are progeny from a line of rats who continually did better in mazes while the other rats distributed to the student were just normal rats not breed from this line of excellent maze runners and they will continue to breed rats that do exceedingly worse in the maze. "The students tested the rats, and confirmed the effects of selection: the bright ones produced good performances in learning while dull ones performed rather poorly (Despret 2004, 117). In reality all the rats were the same, there were no "supergenius gene strand of maze running rats and no strand of slacker maze meanderers. What the experiment showed was how human expectations can be projected onto, in this case, another animal. A point Despret thinks Rosenthal and his students missed is how the rats were living up to expectations and learning what those expectations were. She suggests both rats and humans were shaping each other. They were entering into a relationship in which the "rat proposes to the student, while the student proposes to the rat, a new manner of becoming together, which provides new identities: rats giving to students the chance of 'being a good experimenter', students giving their rats a chance to add new meanings to 'being-with-a-human', a chance to disclose new forms of 'being together'" (Despret 2004, 122). In another article, "Cosmoecological sheep and the Arts of Living on a Damaged Planet," Despret shares similar thoughts in discussing relationships between herders and sheep when she writes: "Thinking about our life and behavior in distinct societies not as disturbance but as integrated parts of systems has great implications. We are invited to pay attention to the health of ecosystems from the inside" (Despret 2016b, 26). Both her idea of an anthropo-zoo-genetic practice and paying attention to ecosystems from the inside bare interesting resemblances to Barad's intra-actions and entanglements, even though I cannot find any citation in which one cites the other's work. Both share a commitment to understanding how humans intra-act with other species and beings and how those intra-actions shape what we refer to as environments, ecosystems, and universes. It should be no surprise then that both scholars are searching for a different way to think about what it means to do science.

A second dimension of Despret's attempt to create a different way to do science is to build from Donna Haraway's work. If we are to think differently about how science is done we will have to find a way to honor what Haraway refers to as "companion species." The dominate mode of relationships with non-human animals is "the act of killing" and this act is covered over and numbed by the sheer number of killings, that in turn becomes data. "Data," Despret (2016a, b, 85) forcefully asserts, "eventually has a role similar to that of the logic of the sarcophagus: to prevent thinking, to forget." How do we avoid not thinking? Not killing or eating non-human animals may be the answer for some but cannot be the answer for most. How do we honor those who are slaughtered in the mass killing animal industries but also those in laboratories and other research environments? "This way of honoring," Despret (2016a, b, 86) concludes, "still remains to be invented. This invention requires that we pay attention to words, to the ways of saying that validate ways of acting and being; it requires us to hesitate, to invent new tropes…" This seems to me to go a long way to move in the direction of honoring and Despret is at the forefront of this process. It also seems to me that Despret's statement above demonstrates her humility because in every aspect of her work from critiquing science protocol to reclaiming alternative forms of knowledge she is searching for and revealing ways to honor non-human animals in science and beyond.

There is yet a another way in which Despret seeks to understand the relationship between non-human animals and science. Like Barad, Despret is interested in rethinking the notion of objectivity. In her article "Becomings of Subjectivity in Animal worlds, Despret (2008, 125) suggests with objectivity "everything is constructed in such a fashion as to render the researcher as impersonal as possible, to make of him a being replaceable by whoever, which is precisely the contrary of what defines a person." In calling out the impersonal, interchangeable, insert scientists or subject here, approach to doing science Despret asks if this undermines any attempt to generalize from a specific study or experiment? "Generalization is always possible," Despret (2008, 128) concedes, "but it is construed in another way; it is constructed bit by bit." This approach she believes requires a different "we". Despret admits she "was feeling more and more acutely irritated in hearing philosophers, psychologists, sociologists, and anthropologists define, with crude generalization, the specificity of humans and their differences from animals. There was clearly a double 'we' in this story,…first, a 'we know full well' and then, a 'we are different'. How could this 'we' impose itself so self-evidently if not as the effect of an academic strong-arm tactic?" In other words, "we" know what is best for non-human animals, if they feel or not, if they are instinctual or thoughtful, if they live or die, and if "you" question our "we", we will call you out as not academic, unreliable, and merely users of anecdotes. Power plays indeed. In place of this strong arm "we", Despret wishes to create an animated subject. She wishes to give life to the subject. Often subjects are construed without agency. "Anonymity, this unquestioned condition of most research investigations," Despret (2008, 130) suggests, "is not simply a characteristic of the research process. From the outset it translates a certain type of relationship and a certain manner of defining those whom one addresses…it suggests a very specific kind of subjectivation." This subjectivation produces "a radical asymmetry of expertises: on the one hand there is the 'researcher-author- the author of questions, of interpretations of hypotheses, of constructions of problems; on the other hand, there is a social actor: witness, informant, someone having opinions…For instance, scientists cite their colleagues by name, whereas the witnesses they talk to are all anonymous"… These anonymous informants are required to "submit to questions, submit to inevitable play of interpretations that will judge one's testimony…The subject is summonsed by a problem that he or she often has nothing to do with, or in any case has nothing to do with the manner in which the problem is defined" (Despret 2008, 131). The subject need not worry about anything just submit, submit to the scientists who know what is important and what is truth. In this sense human subjects share the same position of submission as non-human animals. One way to give life to the (non) human animal subject is to erase the stigma of anecdotes and enter into a more open relationship with (non) human animal subjects that do not present closed notions of objectivity, generalization, and other aspects of research protocol. The honor of (non) human animals and the meaning of science is at stake.

8.4   Women Who Make a Fuss!

Traweek, Barad, and Despret have all recognized the risks they have taken in challenging the unwritten rules, protocol, and truths of science. Traweek and Despret acknowledged the price they have had to pay for challenging tradition. Barad has not. Is it because she is an insider? She is a particle physicists and wrote about the philosophical implications of Bohr's physics-philosophy. Is it not true that in the hierarchy of knowledge that still exists that philosophers would be accepting of a particle physicist who clearly is higher on the pyramid? Traweek is an anthropologist and Despret a philosopher whose original focus was psychology, clearly two fields low on the pyramid—somewhere between biology and history but certainly not as low as education. Traweek had to gain the blessing of the physicists as she pointed out for her work to be legitimated. If she did not apparently her field and department would not have been willing to acknowledge her credentials as worthy of tenure. Her experiences are an example of mental monogamy discussed earlier in this chapter. Traweek (1992, 434) writes academics "assume that there is only one way to think and write carefully and precisely about social and phenomenal worlds." Why are so many academics conservative mental monogamists?

Despret's rejection in the academic world was enough for her to join with Isabella Stengers, who shared similar experiences, to write *Women who make a fuss: The unfaithful daughers of Virginia* Woolf. They ask, in spite of stellar credentials, "why is neither one of us a full professor?" (Stengers and Despret 2014, 13). Could the university not be as open as professors think it is? There were clear limitations to their acceptance into the fraternity called philosophy and the university. "You are welcome and your presence is normal," Stengers and Despret (2014, 17) write in the voice of traditional male philosophers, "for we are 'democrats,' but on our terms, so that nothing changes. You are welcome as long as you do not make a fuss...." Do what you are told, publish in the right journals, cover the right topics, pretend your ideas are original (written in the right "style?", the right format, following the right steps (five sections in all articles with clear and concise sentences, and developing the right conclusions), attend to all your committee assignments (service is important you know), and pay reverence to all senior faculty who by right and privilege can haze and harass you because after all they earned it. Above all do not become the campus radical, you do not want that reputation. Do nothing to make those who will judge you suspicious of your potential. This is what it means to be open and democratic on a university campus? Stengers and Despret (2014, 26) in the midst of this depressing institution write: "What is this civilization where appeals are made to culture and intellectual freedom as if they could stop wars, but whose institutions responsible for cultivating and transmitting culture and free thought work as assembly lines producing beings that Woolf describes as both submissive and violent, thirsting, if not for money, then for recognition, ready for all sorts of brutality...*Think we must*." What kind of scholar thrives on perpetuating such an university? Could it be that the university is not, nor never has been, an institution accepting of new ideas? Is the university a place where only cemented protocols and stagnate traditions thrive, and revolutionary ideas and actions are rejected blindly and unequivocally? Is there any wonder that Freud, Marx, Derrida, Rushdie, Whitman, and, of course, Woolf developed their ideas outside of the university? Yet, if not the university where can transformative thought be nurtured? Stengers and Despret (2014, 23) note that the university is stagnating today because of the outside forces of neoliberalism and its fundamentalist worship of "free markets" at the expense of democratic structures and ideals.

In spite of these conditions they have challenged the protocols and traditions of philosophy and science. They challenged the norms of philosophy "because we imagined that philosophy should be done this way...because that is what we hoped it would be: it was in doing philosophy this way that we showed that it was possible" (Stengers and Despret 2014, 15). They practiced a certain type of philosophy because they believed this is what philosophers were supposed to do and in practicing their philosophy they demonstrated it was possible to think differently and ask different questions in order to violate artificially constructed disciplinary boundaries. In becoming violators of borders, they became interlopers of science with the conviction that "'true science' is recognized by the way in which it destroys our illusions and makes us face truths that we must accept" (Stengers and Despret 2014, 40). In making a fuss they changed the monologues and the routines into controversies and dialogues. "Times have changed," they assert. "The question of a science capable of opening itself to questions that it has traditionally judged 'non-scientific'—including the questions raised by the definition and the requirements of a scientific career and by the formation of future scientists—belongs more than ever to the future" (Stengers and Despret 2014, 44). Yes, they are amateurs, causing intellectual turmoil, overturning protocols masquerading as writing and truths. Yes they are making a fuss, and the university, science, and philosophy owe them an abundance of gratitude.

References

Barad, K. (2007). *Meeting the universe halfway: Quantum physics, and the entanglement of matter and meaning*. Durham: Duke University Press.

Darch, P., Borgman, C., Traweek, S., Cummings, R., Wallis, J., & Sands, A. (2015). What lies beneath?: Knowledge infrastructures in the subseafloor biosphere and beyond. *International Journal of Digital Libraries, 16*, 61–77.

Daston, L., & Galison, P. (2007). *Objectivity*. New York: Zone Books.

Despret, V. (2004). The body we care for: Figures of Anthropo-zoo-genesis. *Body & Society, 10*(2–3), 111–134.

Despret, V. (2008). The becomings of subjectivity in animal worlds. *Subjectivity, 23*, 123–139.

Despret, V. (2015). The Enigma of the Raven. *Angelaki: Journal of the Theoretical Humanities, 20*(2), 57–72.

Despret, V. (2016a). *What would animals say if we asked the right questions?* Minneapolis: University of Minnesota Press.

Despret, V. (2016b). Cosmoecological sheep and the arts of living on a damaged planet. *Environmental Humanities, 8*(1), 24–36.

Latour, B., & Woolgar, S. (1986). *Laboratory life: The construction of scientific facts*. Princeton: Princeton University Press.

Murillo, L. F. R., Gu, D., Guillen, R., Holbrook, J., & Traweek, S. (2013). Partial Perspectives in Astronomy: Gender, Ethnicity, Nationality and Meshworks in Building Images of the Universe and Social Worlds. *Interdisciplinary Science Reviews, 37*(1), 36–50.

Pickering, A. (1992). *Science as practice and culture*. Chicago: University of Chicago Press.

Stengers, I., & Despret, V. (2014). *Women who make a fuss: The unfaithful daughters of Virginia Woolf*. Minneapolis: Univocal Press.

Traweek, S. (1988). *Beamtimes and lifetimes: The world of high energy physicists*. Cambridge, MA: Harvard University Press.

Traweek, S. (1992). Border crossings: Narrative strategies in science studies and among physicists in Tsukuba Science City, Japan. In A. Pickering (Ed.), *Science as practice and culture* (pp. 429–465). Chicago: Chicago University Press.

Traweek, S. (1996). Unity, dyads, triads, quads, and complexity: Cultural choreographies of science. In A. Ross (Ed.), *Science wars* (pp. 139–150). Durham: Duke University Press.

Part V
Interlude Five. Dear David: An Open Letter to a Science Educator

I normally do not write letters because we are no longer a letter writing culture. This is one of the reasons I am writing you to recognize your contribution to science studies and the impact your book, *Procedures of Power and Curriculum Change: Foucault and the Quest for Possibilities in Science Education*, has had. I do not think the field of curriculum studies recognizes the importance of your work and along with Noel and Annette Gough you were an early advocate for the incorporation of science studies into curriculum studies discourses. Not enough people have heard your voice. This is a major problem in our field. Too many are beholden to a star system thereby creating a mono-vocal culture. Curriculum Studies is not the only field that has succumbed to this celestial system, literature has too. The problem with a star system is not that stars exist, but that the star gazers only hear a few voices and drown out other possible voices. I am not a star gazer, but I am grateful for those who arrived before me and opened up intellectual avenues to explore.

In the final chapter of this book I could have selected any number of intellectuals to draw upon from philosophy when discussing science. I could have selected Heidegger who famously rejected the modernist worship of science and reminded his readers that one could not be a good scientist and not be a philosopher. Philosophy was the key to understand science, and therefore more important. I could have selected a more contemporary thinker like Bernard Stiegler whose work explores the ways in which through technology and science life is becoming more monocultural and mechanized. Stiegler does not think technology and science innately leads to an algorithmic reaction to experiences but the way in which technology and science are utilized and developed does. I also could have focused on Michel Serres work. Unlike Heidegger and Stiegler here is a scientist who turns to philosophy because science has become too dangerous to be left alone. In his interview with Bruna Latour, another scholar I could have focused on, Serres (1995, 15) revealed that which underlined his thinking about philosophy and science, and why science cannot be left to the scientists: "Since the atomic bomb, it had become urgent to rethink scientific optimism. I ask my readers to hear the explosion of this problem in every page of my books. Hiroshima remains the sole object of my philosophy." Serres' questioning of scientific optimism remind you of someone else? You too have challenged this optimism without abandoning science. For creating this path I thank you.

All three of these thinkers deserve more attention in matters of science than curriculum scholars are giving. Michel Foucault is one philosopher I could not devote this last chapter to because you already did it. It is not that only one person can write about an important thinker as Foucault nor is it because there is only one possible interpretation to create around Foucault's work and you cornered that market. If this were the case then clearly I have not been reading my Foucault, Derrida, or Nietzsche close enough. The reason I cannot focus on Foucault is I share your reading of his work as it relates to science. You have created an important work, something for curriculum scholars to admire.

Your call to adopt a science and technology studies (STS) approach to science education in order to reform science education is admirable. You are correct I believe when you wrote that adopting an STS approach to learning science was "a chance to debate and consider the role of science and technology in the society they [students] will form and presently influence." (Blades 1997, 36) Science is too important to leave to the scientists and non-scientists in a democracy need to be involved in science research agendas and policies. STS, as you say, science studies, as I say, is an avenue to connect non-science people to the matters of science. This does not mean all people should be in a scientific laboratory one time or another in their lifetimes or they should be involved in every scientific debate that emerges in regards to important questions as to what research should be funded, what policies should be created regarding a scientific discovery, or how scientific creations should be utilized in society. They should however be aware of scientific developments and their impact on human and non-human lives. They should be aware as well of the historical and philosophical meanings of science and how these histories and philosophies impact the lives of billions of people today and tomorrow. Of course, if they so choose they should also have, as a citizen of a democracy, the right to be involved in any scientific matter they deem important. This right does require a responsibility. If an individual thinks they need to be involved in a debate over a scientific development or policy decision, then they must take the responsibility to educate themselves. This is where your work would serve everyone well. It would be in public education science classes where individuals would learn what the scientific debates and policy implications are regarding these debates.

Although you focus on the work of Foucault you are very much a Nietzschean dancer. You are not a sober thinker Nietzsche warned us about. You revel in the beauty and creativity of science. Need I remind you of what you wrote in 1997? It is rhetorical question so of course I do. When reflecting on your experiences of implementing a new science curriculum in Alberta you recognized that curriculum reform and "research," real reform not neoliberal ideological changes in public education, "is a messy, personal business. Clear steps will not be obvious and no product forthcoming. I can not tell someone how to begin or proceed, neither do I recommend anyone follow in the footsteps I have taken, even if I could retrace them with perfect clarity." (Blades 1997, 97–98) Yes! Recipes, steps, protocols, methods kill the spirit of inquiry and cover up the evidence of human endeavor, scientific or not. Do we as educators wish to be accessories to the crime of rewriting the past, covering our chaotic tracks, and pretending we knew everything beforehand? Or do we wish to embark on a quest to create something worthwhile and meaningful? If we choose the latter as you have, then we have to accept what you have: there are no shortcuts, bullet points, or crib notes. There is only the hard work of creativity.

It is not an accident that I just used the word quest because this is what you did in your book. When I re-read your work I kept on saying to myself David's quest is Zarathustrian. He is Zarathustra descending from the mountain looking for his people; the new people who would transcend their humanness and think differently. In your quest you came upon Foucault. No surprise to me that your paths would cross. You "recognized him from a picture on one of his books. No one else in your party paid any attention to him and simply walked by, but I knew from his writings that this scholar had spent time in the Lonely Mountains at the edge of the Kingdom." (Blades 1997, 133) Foucault was Zarathustrian as well. He did spend time in solitude on his own mountain, trying to descend it numerous times to help other people realize their predicaments in life. Did he fail like Zarathustra and return to his mountain lair where he rather spend eternity alone than with people who were unwilling to claim their will to power? Are you? I am not sure. Foucault can no longer answer such queries and only you can answer my question. I do know that your use of allegory was an effective rhetorical move to not only argue for the need for science reform and curriculum debate, but you showed what such a journey can mean to an individual and society if science matters are studied carefully. Thank you for sharing your journey and creating a path. It is a path I started on and then created my own. Yours is why I decided to focus on Nietzsche in this last chapter of my book and why I decided to do it in the form of aphorisms.

References

Blades, D. (1997). *Procedures of power & curriculum change: Foucault and the quest for possibilities in science education*. New York: Peter Lang Publishers.

Serres, M with Latour, B. (1995). *Conversations on science, culture, and time*. Ann Arbor: University of Michigan Press.

Chapter 9
Nietzsche's Science

To write about Nietzsche and science is risky. It is too easy to lump Nietzsche, and those who write in support of his views, with the anti-science groups who use trite and meaningless phrases today like "I don't believe in Evolution," "Evolution is just a theory," or "I don't believe climate change is caused by humans." These knee jerk contrarians, backed by organizations and corporations with profit driven agendas, act as if science is an opinion poll and if enough humans say it is not true then well it must not be true. It must be just those pesky scientists pushing their radical agenda of facts, data, theories, and worldviews. It is risky times to write about science because so many people are willfully misinformed and miseducated about science matters. Nietzsche presents a bold and stark vision of what science is and could be. He shared his own criticism of scientists, those sober thinkers, and the science, utilitarian ordering, they practiced in the name of truth. To group Nietzsche with the anti-science voices of today is to ignore the important insights Nietzsche developed in his life time about science. In my mind there is no doubt that if science were practiced as Nietzsche prescribed and envisioned it would be a more vibrant intellectual endeavor. It would be more poetic and artistic. It would be a gay science, filled with intellectual dancing, boundless intoxication, and rhythmic chaos; just as nature is. Nietzsche's philosophy of science is exactly how scientists should respond to the anti-science movement in the USA. But to do this scientists will have to embrace a few unpleasantries about their own practices and the blindness of their own presuppositions. The path to Nietzschean science is not made easy, but it can be worth the effort. That is what I hope to show in this chapter.

I am going to present this chapter in aphorisms just as Nietzsche did for most of his writings. Before I present the aphorisms I want to establish a few foundational points concerning the importance and uniqueness of Nietzsche's philosophy of science. First, Nietzsche viewed science as a human endeavor but it did not reveal the laws of nature. For Nietzsche like human life, anything in "nature", the earth, and the universe was chaotic. There simply is no meaning in the world. The only meaning that existed was invented by humans. There is also no truth, just human interpretations. This may sound like anti-science opinion poll statements like "I

© Springer International Publishing AG, part of Springer Nature 2018
J. A. Weaver, *Science, Democracy, and Curriculum Studies*, Critical Studies of Education 8, https://doi.org/10.1007/978-3-319-93840-0_9

don't believe in..." but it is not. It could not be any further from the anti-science flat(head) earthers of today. Nietzsche without a doubt would call out the anti-science voices of today for who they are; guilt ridden Christians (and some Muslims too) driven by self-pity, ressentiment, and rage. Nietzsche in place of this herd mentality of "I don't believe..." statements wanted to create what Babette Babich refers to as perspectivalism or the creation of human interpretations that are the best possible human creations at the time and whose interpretation will be accepted until something better is created by humans to explain some aspect of life, in all its chaos. Perspectivalism recognizes that human creation is all we have but there are interpretations that are better than the rest. This is what Nietzsche sought. As usual, though, there are always caveats with Nietzsche. This best interpretation was not the one that tamed nature and subsumed it under some (human) universal laws. There is no taming of nature or universal laws, just human creations. The best interpretation for Nietzsche needs to recognize the human role in human science. Nietzsche's demand that the human role in inventing the universe be recognized did not sit well with many of his contemporary scientists, especially those who represented a positivistic view of science, nor do his views sit well with many scientists today. However, it does not matter what scientists think. They can hide their presuppositions all they want from themselves, but Nietzsche still sees the presuppositions thereby rendering any proclamation of truth about nature or the universe a human invention.

Just because scientists are at odds today with the anti-science pious ones, does not mean scientists are immune from the sickness that plagues the pious. For Nietzsche, he did not see much difference between the religious and scientific. Both suffered from an ascetic ideal. The ascetic ideal which can be found in St. Paul, most religious leaders today, certainly politicians of pity, and too many intellectuals and scholars, is an ideal opposed to life. Life for Nietzsche in all its glory was not about Mill's and Bentham's philosophy of pleasure over pain and the greatest good for the greatest number of people. This is merely the philosophical embodiment of what ails the world today, it certainly plagues USA politics and religion. The ascetic ideal rejects the reality of life; a life of meaninglessness, misery, and pain. It takes this rejection of life and replaces it with guilt, responsibility, and, most importantly pity. The anti-science pulpit preachers govern their realm of anti-life through the use of pity. Their faith is so weak they shame parishioners into believing that one cannot have a religious faith and believe that humans are a major cause of global warming or that evolution is indeed a very effective perspective of science, the best we have at this moment. From this ascetic ideal grows an outrage against anyone who might think differently, or simply just thinks. The ascetic ideal is an embodiment of a negative nihilism in which life despises life. As Deleuze (1983, 119) notes this ressentiment sets up a "formula" that "must be compared with that of the master: I am good, therefore you are evil." Rage defines the pious today in the religious communities of the USA.

Scientists suffer from the same rage at times. Scientists however do not use the pity of god as USA Christians do. They instead use the club of "truth" to justify their rage; their tempered, "objective," and "value neutral" rage. "Oh you reject my, which is not mine but nature's, truth; a truth I, not I but nature, constructed from a model, a formula, a system, a theory, and then denied that my model, formula, system, or theory had anything to do with this truth." This is a form of negative nihilism, it uses guilt, pity, and rage to force someone to think the "right" way. A more positive form of nihilism which Nietzsche embodies would be to recognize that scientists construct models, create systems, develop formulas, and craft theories in order to provide a human sense of order over the reality of a chaotic world. The truth of the proclamation of owning a truth claim would not be the truth claim but it would be in the proclamation that it is a human truth claim. To proclaim a human construct something all too human would be a statement in the name of life; chaotic, open, miserable, glorious life. If scientists cannot see their own human constructions then they are merely another embodiment of the ascetic ideal that has plagued the world for millennium. Science under an ascetic ideal becomes a will to truth not a will to power or as Arthur Danto (1965, 227) writes, it "is the teaching that the world is something we have made." The will to truth, on the other hand, is interested in making impossible, absolute claims that are pronounced as iron clad laws, universal truths, set protocols and formulas which says nothing about reality but merely references what Rebecca Bamford (2005) calls "the illusion of truth itself." In this regards, the anti-science ascetes are onto something. Evolution is just a theory, a will to power. Yet again pity takes over quickly and we find the ascetes are back into their rage mode because evolution is the best we have at the moment and a good example of human will to power. Pity the religiously pious ones because evolution is willing itself into the science classroom and the religious freedom of many to be part of the ascetic herd is being denied. Evolution as a will to power, however, does not mean someone like Louis Agassiz or Herbert Spenser were correct to use science to rationalize their own form of racism and colonial repression. Scientific racism is not a form of will to power. It is a will to truth in which the name of science is used to rationalize slavery and colonial repression. It is a form of the ascetic ideal and how science can be no different than religion when it is abused by the faithful.

Nietzsche sought something more than the ascetic ideal. He sought life in all its glory, reverence, misery, poverty, and potential greatness; life in all its possibility. Nietzsche wanted a science that would not hide its presuppositions or its agenda but would assist humans in creating meaning out of the chaos of the universe. This could not happen unless scientists recognized they were partaking in a human endeavor and since this was a human endeavor it meant that as scientists they were artists. Not art as in scientists need to start reading poetry and take up painting, art as in the idea that "science as a method of inquiry or investigation and science as a worldview or world-construction" cannot understand itself. To be an art means to accept that "science is not only the investigation but also a representation of nature: the discovery and expression of facts" (Babich 1994, 37). Art is understanding the limits of science and creating a world out of these limits and the chaos that engulfs the world.

9.1   Aphorisms on a Nietzschean Science

57

> *To the realists.*—You sober people who feel well armed against passion and fantasies and would like to turn your emptiness into a matter of pride and an ornament: you call yourselves realists and hint that the world really is the way it appears to you. As if reality stood unveiled before you only, and yourselves were perhaps the best part of it. (Nietzsche 1974, 121)

1

> To read Nietzsche, as to play or to perform a musical piece, is to interpret Nietzsche: and each time that interpretation must differ. The achievement, the interpretation of the text testifies as much to Nietzsche's philosophic artistry as to the reader's. (Babich 1994, 30)

1

How are we to be the dancers of science Nietzsche wishes we become; interpreters of chaos, courageous explorers of everything and inquisitors of life, when the sober people insist we follow protocol and most definitely abandon our joy and rhythm to dance to the beats of what we might call all that is nature? We cannot ignore the sober scientists, they are important as our guides and teachers. But what are they teaching us if they are telling us not to dance and to tame our passions and fantasies? We need guides and teachers who learn to dance again, commune with their muse called science and nature but these dancers should not think falsely that somehow we humans are the center of nature and they, the scientists, are the only ones who know how to unlock the mysteries of the dances of nature. The only thing they are unlocking are set ways, self-pity, and blind followers of a system disguised as truth and abused as the name of truth. Abusers of truth are not scientists but ideologues, priestly predators.

1

> All this is to say, the human achieves the blissful feeling of existence in two states: in dreams and intoxication. The beautiful seeming...of the dream world, in which every person is the consummate artist, is the father of all the imagistic arts...Dionysian art is centered on the play with intoxication, with the state of ecstasy...The festivals of Dionysus not only forge a union between man and man, but reconcile man and nature. (Nietzsche 2013, 29 and 31)

2

Intoxicated with what? Baudelaire (1988, 55) said it best almost two centuries ago "With wine, with poetry, or with virtue, as you choose. But get Drunk." Become intoxicated with science and explore not to be like the sober people, but to see the world anew; create a new interpretation of what the chaos might be. Create your own intoxicated sobriety, remain restless until your science or understanding of science becomes an art, and the illusion of human and nature separated is exposed as an invention of the sober people who forgot among many things that humans and nature were never apart just interpreted as such.

123

> *Knowledge as more than a mere means*—Without this new passion—I mean the passion to know—science would still be promoted: after all, science has grown and matured without it until now. The good faith in science, the prejudice in its favor that dominates the modern state (and formerly dominated even the church) is actually based on the fact that this unconditional urge and passion has manifested itself so rarely and that science is considered not a passion but a mere condition or an 'ethos'…for many people it is actually quite enough that they have too much leisure and do not know what to do with it except to read, collect, arrange, observe, and recount—their 'scientific impulse' is their boredom. (Nietzsche 1974, 178–179)

3

Dancing, then, is not enough. We have to reclaim or as Nietzsche suggests claim, science in the name of passion. We may even have to wrest science away from those who promote a passionless science. The passionless ones will tout their accomplishments and awards not as symbols of creativity and intoxication but as weapons of power. They will proclaim the dancers, the passionate ones, the artists know nothing of science and should let the readers, collectors, arrangers, observers, and counters alone; allow them to do their seriously sober work: the economy depends on them, sick people need to be cured, the earth saved, and the truth known. When the time comes they will let us know when we can perform and speak, they will tell us what to do and how it should be done. These power plays should not be heard, but if they are, laugh; not out of disrespect but out of respect. Sickness, poverty, well-being, truth and the earth are too important to listen to those who wish to limit our abilities to create interpretations of what we can do to address the problems humans, non-human other animals, and the earth face.

324

> *In media vita"*…'Life as a means to knowledge'—with this principle in one's heart one can live not only boldly but even gaily, and laugh gaily, too. And who knows how to laugh anyway and live well if he does not first know a good deal about war and victory? (Nietzsche 1974, 255)

3

One repays a teacher badly if one remains only a pupil. (Nietzsche 1969, 103)

2

> No learning can avoid the voyage. Under the supervision of a guide, education pushes one to the outside. Depart: go forth. Leave the womb of your mother, the crib, the shadow cast by your father's house and landscapes of your childhood. (Serres 1997, 8)

4

Repay your teachers well. They have created an understanding of the world, the sciences, nature. They, mostly, have only forgotten that their creations are creations, their universal truths their presuppositions about universal truths, their models are models and not nature. Intoxicate the sober ones but allow them to create freely as well, perhaps soon they will awake from their slumber and use their artistic abilities to create even more impressive worlds. But this time, let us hope they will be more open and honest about their human roles in creating their human sciences. Remind them what the foundation of their sciences is.

115

> *The four errors.*—Man has been educated by his errors. First, he always saw himself only incompletely; second, he endowed himself with fictitious attributes; third, he placed himself in a false order of rank in relation to animals and nature; fourth, he invented ever new tables of goods and always accepted them for a time as eternal and unconditional: as a result of this, now one and now another human impulse and state held first place and was ennobled because it was esteemed so highly. If we removed the effects of these four errors, we should also remove humanity, humanness, and 'human dignity'. (Nietzsche 1974, 174)

3

> It is not my intention, nor was it Nietzsche's, to denounce the illusion that is the scientific project. For, according to Nietzsche, illusion is necessary for life itself...Logic simply offers one aesthetic possibility or perspectival valuation among others....The scientific world is a human construction. Scientific reality is an illusion, it is not the Real as such, and it is not true...This illusion of scientific truth, for Nietzsche, is that one can explain reality when one has merely given it a mathematical and logical...description. (Babich 1994, 146–147)

5

Just as art is a foundation for science so is error. It is an error to construct an ordered hierarchy in which humans are given dominion over the earth and other non-human animals are somehow expendable and not nearly as important. It is an error of our collective illusions called science, but the point is not to call out these errors. It is our task as scientists and non-scientists to create newer and hopefully better illusions. The illusion of perfect order, neat classifications, universal man is being undressed, we are in a time of collective psychoanalysis in order to rethink the consequences of our illusion. Some are in denial and lashing out with pity and rage, continuing the tradition of the ascetic ideal, but many more are admitting to the wounds caused by the errors of science. Those of us who are confessing are ready to create some new errors and accept a new illusion. What will we create? How will we dance around the errors we dressed up in truth this time? Will our moves be healthier choices for the earth, humans? Will species survive or will the dominionists, the self-pity righteous ones prevail? We humans are in a race of creativity. Will we create a new illusion that allows non-humans and the earth to thrive?

58

Only as creators!—This has given me the greatest trouble and still does: to realize that what things are called is incomparably more important than what they are. The reputation, name, and appearance, the usual measure and weight of a thing, what it counts for—originally almost always wrong and arbitrary, thrown over things like a dress and altogether foreign to their nature their nature and even to their skin—all this grows from generation unto generation, merely because people believe in it, until it gradually grows to be part of the thing and turns into its very body…. We can destroy only as creators.—But let us not forget this either: it is enough to create new names and estimations and probabilities in order to create in the long run new 'things.' (Nietzsche 1974, 121–122)

6

The pious who live their lives in pity and rage know that naming something is merely enough. The White House is now occupied by a chiefly pious one. Say anything and it shall be labeled as fact and so it shall be, said the gods of deceit and corruption. These pious ones though Nietzsche did not have in mind. They are the negators of life not the affirmers. Nietzsche was interested in those artists, scientists or not, who could create new worlds and then name those worlds so others could live in those worlds nobly. Galileo was a creator of new names, new moons of Jupiter, new thinking of the earth but the pious ones controlled him even after his death. Newton invented such a novel system the most learned of people thought he was making it up. Newtonians literally had to be educated and nurtured until anyone else understood what Newton was trying to create. Climate scientists have spent six, seven, ten years of their lives creating computer models in order to create the data to understand climate change. Will we listen to these destroyers or will we embrace the ragging pious deniers who dress up their corporatized agendas in scientific dress to challenge the warming of the planet, the destruction of species, the disappearance of island nations and coast lines? What will new creators destroy in order to make science an art again? Will we embrace the destruction or destroy the creators in the name of utility, profit, luxury, and method?

373

'Science' as a prejudice.—…It is no different with the faith with which so many materialistic natural scientists rest content nowadays, the faith in a world that is supposed to have its equivalent and its measure in human thought and human valuations—a 'world of truth' that can be mastered completely and forever with the aid of our square little reason. What? Do we really want to permit existence to be degraded for us like this—reduced a mere exercise for a calculator and an indoor diversion for mathematicians?…that the only justifiable interpretation of the world should be one in which you are justified because one can continue to work and do research scientifically in your sense…and interpretation that permits counting, calculating, weighing, seeing, and touching, and nothing more—. (Nietzsche 1974, 334–335)

4

For Nietzsche…the scientific project is the discovery of the Real as calculable. The *Objectivity* of this ratiocinative, calculative interest is not the equivalent to innocence; nor does such an objective focus (alone) justify the scientific project…There is no interest-free knowledge project, and science itself, in its logicizing and mathematizing…has a special interest in a fixed image of the world. (Babich 1994, 140)

7

To quantify the world was once considered a socialist, radical endeavor. It always was a state function. Now it is a function of biopolitics to determine life, death, war, actuary tables, and disease. But it should never have been viewed as a means to tame the creators. Dancing scientists have to find ways to break the monopoly of calculation, quantification, measuring, and ordering; all destroyers of life, weapons of the pious, and language of the sober people. Yet, it is not calculation, quantification, measuring and ordering in itself that sobers creativity. It is the attempt of sober people to monopolize science and define their way as the only way; it is the capitalists' assumption that anything of value must have a calculable value, a utility. This is what is strangling educational research. The monopoly of quantitative "research" and that research of its step-child, qualitative research as a conspiratorial binary ("I won't reveal your illusory nature if you don't expose mine"). Too many voices are eliminated by this binary and the myopia of the sober people. It is not just in educational research but in all of life. Utility is killing life in the name of profit margins.

15

If we wish to consider Socrates as one of these charioteers, we need only see him as the prototype of a new and unimagined life-form, the prototype of theoretical man. Our next task must be to understand the significance and purpose of that figure. Like the artist, theoretical man takes an infinite delight in everything that exists, and, like him, he is shielded by that delight from the practical ethics of pessimism with its eyes of Lynceus that glow only in the dark. Whenever the truth is uncovered , the artists gazes enraptured at whatever covering remains, but theoretical man takes delight and satisfaction in the covering that has been cast aside, and takes greatest delight in a process of uncovering that is always successful and always achieved by his own efforts. (Nietzsche 1993, 72)

18

The whole of our modern world is caught up in the net of Alexandrian culture, and its ideal is theoretical man…All of our educational methods take their bearings from this ideal: any other form of existence has a hard struggle to survive alongside it, and is in the end tolerated rather than encouraged. In an almost frightening sense, the man of culture has long existed only in the form of the scholar… (Nietzsche 1993, 86)

5

For some reason it was the fate of America not to become itself, not to build its house upon the foundation of a loss for which no recovery was possible. Its fate, rather, was to sacrifice its freedom to nationhood, to reiterate and exasperate the rage for possession, and to fall into the watery mire of what is not life. Instead of a nation of poets, it became a nation of debtors, property owners, shopkeepers, spectators, gossipers, traffickers in rumor, prejudice, and information—capitalists who in their strange uncertainty about life pursue the delusions of recovery for their appropriation of everything. (Harrison 1992, 231)

8

Who shall we become as Curriculum Scholars? Shall we become the scholars of sober people? Will we become the Nietzschean theoretician who shares the same curiosity as the artist but very different outcomes? The artist revels in life to create life, the theoretician marvels at mystery in order to unravel, expose, and systematize it into some formula, system, or protocol; proclaiming all the way that "I did that, I invented that system." Will we become, like so many theoretical people, slaves of neoliberalism, servants to capitalism? Will we seek out the latest grant stream of income so our university can claim us and our "research" while we are being paid by some corporate entity that wants a specific kind of profitable and marketable result so it can share the patent with the university and the theoretical person can become an entrepreneur—the god and goddess of a nation of debtors and traffickers of prejudice? If curriculum scholars say no to this theoretical people, no to neoliberal capitalism, and the nation of gossipers, then we cannot remain aloof to the demands of science. To ignore science matters and Nietzsche's call to become a destroyer, a dancer, a life affirming philosopher is to become complicit in neoliberalism and intellectual sobriety; to become a shopkeeper of your own little fiefdom. Robert Harrison (1992, 238) ended his little commentary on what the USA became after Emerson and Whitman by suggesting that "when a nation loses its poets it loses access to the meaning of dwelling. When it loses the meaning of dwelling, it loses the means to build…for when a nation ignores its poets it becomes a nation of the homeless." I would add when a nation does not wed science to art, and the artists, including the poets and curriculum scholars, and ignores science it becomes a nation of profiteers blinded by its superficiality, cowardly narcissism, and navel gazing nihilism.

21

> *To the teachers of selfishness.*—…(The most industrious of all ages—ours—does not know how to make anything of all its industriousness and money, except always still more money and still more industriousness; for it requires more genius to spend than to acquire.—Well, we shall have our 'grandchildren'!)
>
> If this education succeeds, then every virtue of an individual is a public utility and a private disadvantage, measured against the supreme private goal—probably some impoverishment of the spirit and the senses or even a premature decline. (Nietzsche 1974, 94)

6

> From early in the eighteenth century, nonconformist circles, particularly in the Midlands and North, supported dissenting academies deeply committed to 'modern' education, including the natural sciences. Later in the century, provincial scientific societies emerged strongly associated with the mercantile and manufacturing classes. We now have a generation of research in the social history of science that establishes significant links between approval of and participation in scientific culture and the emergence of the 'new men' of industrializing Britain. (Shapin 2010, 179)

9

As Steven Shapin's quote above demonstrates science has always had connection to "industry" or business. Science has always been viewed as a discipline of utility, even when it was referred to as a natural philosophy. Chemistry was invented in part to meet the needs of mining and later became essential in the textile and dye industries. Today we just refer to chemistry in part as the chemical industry. Botany was created to help states become self-sufficient, now it is seen as a branch of pharmaceutical corporations in their lust for "natural remedies" in the name of profit. Biology is a little different in its break from natural philosophy, but once genetics became a driving force in the field the race for patents and profit is matched by no other field of science. The transfer of knowledge from science and technology fields within universities to established corporations or faculty led start-up companies is done with great speed today. Start-up companies are the new and preferred means to riches for academics in the contemporary university. Yet must every Goddamn thing, including humans, become utility? Is there anything of value in science that does not have a utility value? Is science just another name for economics? I hope not. Yet this is exactly what Nietzsche is criticizing in his aphorism, he even refers to this reduction of everything to utility as an education. Is this not what we have done to "our grandchildren"—turned them into economic objects? In the United States Christian fundamentalism is thriving at the same time the reduction of everything to utility is complete. Is it not a surprise, these two spiritually deprived movements have destroyed the spirit of humans? Nietzsche was correct to warn us that Science too is not immune from the pious ascetic ideal. Science in its lust for utility has become the embodiment of the death spiral of ressentiment and nihilism. Science, like public schools and religion, has educated their "grandchildren", but it has been a colossal failure for the spirit and well-being of humans and non-human other animals who have long suffered from the utility of capitalist thought. Instead science has educated its "grandchildren" to control nature, tame it, and turn its chaos into certainty (into illusion).

335

The origin of our concept of 'knowledge.'—...Look, isn't our need for knowledge precisely this need for the familiar, the will to uncover under everything strange, unusual, and questionable something that no longer disturbs us? Is it not the *instinct of fear* that bids us to know? And is this the jubilation of those who attain knowledge not the jubilation over the restoration of a sense of security? (Nietzsche 1974, 300–301)

10

The fear of uncertainty has killed the spirit just as the need for accountability via standardized tests has killed the public schools in the USA. Utility deforms the natural, converts the strange into something of human (material) value. Our fear is a great motivator but a terrifying teacher. We should not honor the scientists who bring us security, the sober destroyers of the imagination, but honor those who keep us in mystery and remind us of the complexity and uncertainty of life. We should honor the scientist/artist. This, then, is what I want from science and the scientists: creativity, uncertainty, intoxication, and dance. I want art! I want all to say yes to life.

7

Nietzsche asks us to think the question of science from the viewpoint of the artist, which is not to say that he proposes to frame the idea of science from the cultural imaginary of the social sphere of art but much rather from the viewpoint of art or techne or poiesis as inventive, creative human activity. (Babich 1999, 5)

8

For as philosophers we do need to know what in us wants the truth, what wants scientific regularity, technological mastery. Knowing that about ourselves, as philosophers of this most dangerous 'perhaps,' we are not automatically healed or redeemed. But where more lies in the question mark we place after ourselves than in any other insight, we raise the saving virtue, the grace of suspicion. It is the modesty that is the precondition for heroic valor, the very ironic, tragic meaning of the Delphic 'know thyself.' This is science on the ground of art—in the service of life. (Babich 1994, 298)

12

On the aim of science.—What? The aim of science should be to give men as much pleasure and as little displeasure as possible? But what if pleasure and displeasure were so tied together that whoever wanted to have as much as possible of one must also have as much as possible of the other—that whoever wanted to learn to 'jubilate up to the heavens' would also have to be prepared for 'depression unto death'? (Nietzsche 1974, 85)

11

This is the challenge science presents us. We, curriculum scholars, are not scientists nor from science, but we need to belong to and live in science in order to accept Nietzsche's challenge of jubilating in heaven and death. Will we dance or sit with our hands under our legs not knowing what the rhythm and beats of life are? We need to ask Babich's fundamental question of ourselves and science. Will we, will it, serve the needs of life or utility; uncertainty or security; poiesis or method; intoxication or sobriety? These questions form the ground of curriculum studies as an art in the name of life. Science, though, has to be addressed, named, (dis)honored for our art to serve life.

251

Future of Science.—Science bestows upon him who labours and experiments in it much satisfaction, upon him who *learns* its results very little. As all the important truths of science must gradually become common and every day, however, even this little satisfaction will cease:...But if science provides us with less and less pleasure, and deprives us of more and more pleasure through casting suspicion on the consolations of metaphysics, religion and art, then that mightiest source of joy to which mankind owes almost all its humanity will become impoverished....If this demand of higher culture is not met, then the future course of human evolution can be foretold almost with certainty: interest in truth will cease the less pleasure it gives: because they are associated with pleasure, illusion, error, and fantasy will regain step by step the ground they formally held: the ruination of science, a sinking back into barbarism, will be the immediate consequence; mankind will have to begin again at the weaving of its tapestry. (Nietzsche 1998, 119)

12

How could Nietzsche have predicted our present moment in 1878? Was he psychic? Was he an extraterrestrial just visiting us humans who are blinded by our needs and ideologies? I think not. He was just attuned to life and the reverberations of chaos while others have desperately tried to construct facades to cover over reality. When science broke from conservative, neoliberal norms and proclaimed dire consequences if global warming continued to be ignored, Science no longer was fun. "STEM is fun! Creationism is life affirming, there is a heaven and only the Christian scientists will be allowed in! Global warming is depressing. What can we do, the problem is too big for us. Can't we just continue along our path of the fun-like accumulating more things and driving more cars and on the weekends just recycle more? Will that appease the alarmists? Evolution is evil. Can't we just learn about how the earth is maybe ten thousand years old and that way we all can go to heaven and we can all leave this earth to enter into our constructed paradise? Can't we just listen to the basketball star and accept that the earth is flat? If this does not satisfy the evolutionists, we, Christians, will pray for them." How quickly the tapestry of sanity has unraveled and no matter how insane the ignorance grows in response to the unpleasantries of science, the future grows dim as higher culture leaves even the higher institutions.

References

Babich, B. (1994). *Nietzsche's philosophy of science: Reflecting science on the ground of art and life*. Albany: SUNY Press.

Babich, B. (1999). Truth, art, and life: Nietzsche, epistemology, philosophy of science. In B. Babich & R. S. Cohen (Eds.), *Nietzsche, epistemology, and philosophy of science: Nietzsche and the sciences II* (pp. 1–24). Dordrecht: Springer.

Bamford, R. (2005). Nietzsche, science, and philosophical nihilism. *South African Journal of Philosophy, 24*(4), 241–259.

Baudelaire, C. (1988). Get drunk. In C. Baudelaire (Ed.), *Twenty prose poems* (Vol. 55). San Francisco: City Lights.

Danto, A. (1965). *Nietzsche as philosopher*. New York: Macmillan Company.

Deleuze, G. (1983). *Nietzsche and philosophy*. New York: Columbia University Press.

Harrison, R. (1992). *Forests: The shadow of civilization*. Chicago: University of Chicago Press.

Nietzsche, F. (1969). *Thus spoke Zarathustra* (R. J. Hollingdale, Trans.). London: Penguin.

Nietzsche, F. (1974). *The gay science* (W., Kaufmann, Trans.). New York: Vintage Press.

Nietzsche, F. (1993). *The birth of tragedy* (S. Whiteside, Trans.). London: Penguin.

Nietzsche, F. (1998). *Human, all too human: A book for free spirits* (R. J. Hollingdale, Trans.). Cambridge: Cambridge University Press.

Nietzsche, F. (2013). *The Dionysian vision of the world* (I. J. Allen, Trans.). Minneapolis: Univocal.

Serres, M. (1997). *The troubadour of knowledge* (S. F. Glaser, Trans.). Ann Arbor: University of Michigan Press.

Shapin, S. (2010). *Never pure: Historical studies of science as if it was produced by people with bodies, situated in time, space, culture, and society, and struggling for credibility and authority*. Baltimore: Johns Hopkins University Press.

Printed in the United States
By Bookmasters